AF869622

Springer Water

Series Editor

Andrey Kostianoy, Russian Academy of Sciences, P. P. Shirshov Institute of Oceanology, Moscow, Russia

Editorial Board

Angela Carpenter, School of Earth & Environment, University of Leeds, Leeds, West Yorkshire, UK

Tamim Younos, Green Water-Infrastructure Academy, Blacksburg, VA, USA

Andrea Scozzari, Area della ricera CNR di Pisa, CNR Institute of Geosciences and Earth Resources, Pisa, Italy

Stefano Vignudelli, CNR - Istituto di Biofisica, Pisa, Italy

Alexei Kouraev, LEGOS, Université de Toulouse, TOULOUSE CEDEX 9, France

The book series Springer Water comprises a broad portfolio of multi- and interdisciplinary scientific books, aiming at researchers, students, and everyone interested in water-related science. The series includes peer-reviewed monographs, edited volumes, textbooks, and conference proceedings. Its volumes combine all kinds of water-related research areas, such as: the movement, distribution and quality of freshwater; water resources; the quality and pollution of water and its influence on health; the water industry including drinking water, wastewater, and desalination services and technologies; water history; as well as water management and the governmental, political, developmental, and ethical aspects of water.

More information about this series at https://link.springer.com/bookseries/13419

Aysha Akter

Rainwater Harvesting—Building a Water Smart City

Springer

Aysha Akter
Professor, Department of Civil Engineering
and Head, Department of Water Resources
Engineering
Chittagong University of Engineering
and Technology (CUET)
Chittagong, Bangladesh

ISSN 2364-6934 ISSN 2364-8198 (electronic)
Springer Water
ISBN 978-3-030-94642-5 ISBN 978-3-030-94643-2 (eBook)
https://doi.org/10.1007/978-3-030-94643-2

© The Editor(s) (if applicable) and The Author(s), under exclusive license to Springer Nature Switzerland AG 2022
This work is subject to copyright. All rights are solely and exclusively licensed by the Publisher, whether the whole or part of the material is concerned, specifically the rights of translation, reprinting, reuse of illustrations, recitation, broadcasting, reproduction on microfilms or in any other physical way, and transmission or information storage and retrieval, electronic adaptation, computer software, or by similar or dissimilar methodology now known or hereafter developed.
The use of general descriptive names, registered names, trademarks, service marks, etc. in this publication does not imply, even in the absence of a specific statement, that such names are exempt from the relevant protective laws and regulations and therefore free for general use.
The publisher, the authors and the editors are safe to assume that the advice and information in this book are believed to be true and accurate at the date of publication. Neither the publisher nor the authors or the editors give a warranty, expressed or implied, with respect to the material contained herein or for any errors or omissions that may have been made. The publisher remains neutral with regard to jurisdictional claims in published maps and institutional affiliations.

This Springer imprint is published by the registered company Springer Nature Switzerland AG
The registered company address is: Gewerbestrasse 11, 6330 Cham, Switzerland

Dedicated to

Our daughter Arisha Sadiq

Foreword by Garry Pender

The potential of rainwater harvesting to provide: an independent water supply, supplement the main supply, recharge groundwater sources and mitigate flooding is well known. In fact, the first application of the concept can be traced back to the Neolithic period. Increasing societal pressures arising from floods and droughts in recent years have seen an increased interest in using the technique in both urban and rural settings. Its deployment is a valuable tool in addressing the inequalities faced by the billions of people worldwide who lack adequate access to clean water. Climate changes overstate the cases where predictions indicate that changing geographical and temporal variations in rainfall have the most significant impact on water-stressed areas. In the developing world, there has also been a growing interest in the use of technology. Cities worldwide are seeking novel solutions to make the urban environment more sustainable. Here, rainwater harvesting benefits by reducing demand on water treatment and aging "grey" drainage systems, alleviating flood risk and creating opportunities to "green" cityscapes.

This book is an excellent introduction to recent developments in this topic. It presents a logical and comprehensive introduction to the theory, technology, economics, and implementation of rainwater harvesting. Effectively addresses a potential barrier to the effective use of rainwater harvesting systems by providing a single reference point for engineers, architects, ecologists, planners responsible for their design, implementation, and maintenance. I recommend it to all those studying or implementing these systems.

Garry Pender
Professor, Deputy Principal (Research and Innovation)
Heriot-Watt University
Edinburgh, UK

Foreword by Mukand S. Babel

The World Economic Forum has consistently identified water crises among the top five global challenges for the last eight years, and climate change will only exacerbate the situation. Already, 70% of the world's megacities are water-stressed. The problem is more alarming in developing countries, especially in Asia and Africa, where groundwater levels are depleting rapidly. From experiences around the world, it is clear that there is a need to transition from supply management of water to demand management. I am, therefore, happy to see this book *Rainwater Harvesting—Building a Water Smart City* address one such demand management measure. Rainwater harvesting also helps in addressing urban flooding issues and in enhancing groundwater recharge.

Perhaps the greatest attribute of any smart city is that it needs to be water secure. While the definition of water security may vary from place to place and discipline to discipline, almost all definitions will include safeguarding a sustained water supply for humankind to survive. Given that the water demand in cities will only increase because of rapid urbanization and other development pressures, it will no longer be wise to take conventional water sources for granted. The practice of rainwater harvesting has yielded a positive water balance in cities and is undoubtedly the way forward in this age of deep uncertainty.

The good news is that the enabling environment for rainwater harvesting is quite robust. Several countries have introduced various policies, regulations, and instruments to help scale up this practice. In cities like Delhi and Bangalore, building permits are contingent on mandatory rainwater harvesting installation. In Bangkok, properties are provided with incentives like reduced taxes and subsidies on water bills. Cities like Los Angeles and Melbourne have invested in large-scale community-level rainwater harvesting. All these examples exemplify the potential of rainwater harvesting in addressing and alleviating the water crises significantly.

This book discusses all the relevant aspects of rainwater harvesting that encompasses the conceptualization, design, and implementation of various models; large-scale capture of stormwater; intricacies involved with groundwater recharge; economics of rainwater harvesting in terms of life cycle costs and the cost-benefit ratio; and the technological interventions to help scale up such initiatives.

I congratulate Prof. Aysha Akter on the timely publication of this book. Her experience and expertise in the sector certainly reflect in the book. I am confident that the book will serve as a useful resource for students, city planners, architects, engineers, and practitioners who wish to design and install rainwater harvesting systems in urban areas.

Mukand S. Babel
Professor, Water Engineering
and Management (WEM), Chair,
Climate Change Asia at AIT
(CCA@AIT)
Asian Institute of Technology (AIT)
Pathumthani, Thailand

Foreword by Muhammed Alamgir

It gives me immense pleasure and emotional feelings while I have been requested by one of my beloved former students to write the foreword for a book entitled *Rainwater Harvesting—Building a Water Smart City* written by her, Dr. Aysha Akter, Professor of Civil Engineering, Chittagong University of Engineering & Technology (CUET). At the same time, I became delighted to see that the substance of the book is deeply related to the topics relevant to the country vulnerable to climate change impacts like Bangladesh, which constituted 700 km of coastline covering 47,000 sq.km where 35 millions people are living and facing scarcity of safe drinking water in its coastal zones. The issue is also relevant to achieving the Sustainable Development Goals (SDGs) for ensuring access to water for all.

I have gone through the manuscript *Rainwater Harvesting—Building a Water Smart City*; this seems complete guidance to achieve sustainable goals. Earlier in my life, I have been in Japan for more than five years (1992–1998) as a postgraduate and researcher and resumed my job in Khulna University of Engineering and Technology (KUET), Bangladesh as an academician in 1998. During my stay in Japan, I have seen how they turn a "drain city to rain city" by adopting rainwater harvesting, and this is the best example of "Tokyo lesson". I used to teach the basics of Geotechnical Engineering in the undergraduate classes at the Department of Civil Engineering, KUET. I found a great reflection of those in this manuscript and believe this will provide helpful guidance to the young researchers. Moreover, this book will provide important information to policymakers and professionals, especially planners, architects, engineers, and plumbers, to design and install rainwater harvesting systems in an urban area.

Conventionally, rainwater harvesting has been practicing in Bangladesh for a long time. I have experiences that these rainwater harvesting systems practice rural Bangladesh, especially in the coastal zones, as the primary source of potable water due to salinity and as a supplement source due to arsenic and iron contamination. While in urban areas to support existing water supply systems and available sources to meet the water scarcity and environmental sustainability. In addition, one of the biggest challenges in urbanization is handling municipal solid wastes and greywater. Greywater combined with rainwater could be an excellent means of cutting down urban freshwater usage, which has already been successfully implemented in some Asian countries. This book contains a chapter on the Managed Aquifer Recharges, and this chapter also highlighted how to handle contaminated water from infiltration.

After mainstreaming rainwater harvesting in developed countries' urban areas, this book has compiled the lessons from the existing water smart cities for future city planners and architects. The author of this book has already made a significant contribution in this area to bring research outcomes to protect the planet for the next generation. I am proud to recall one such recognition recorded as she received the prestigious University Grants Commission award 2015 for her outstanding research work in the Engineering Division on "Rainwater Harvesting for an Urban City".

Finally, I hope that the publication of this book will draw the massive attention of the concerned stakeholders throughout the world.

Muhammed Alamgir
Professor, Member, University Grants Commission
Dhaka, Bangladesh

Ex. Vice-Chancellor, Khulna University of Engineering and Technology (KUET)
Khulna, Bangladesh

Preface

Rainwater Harvesting—Building a Water Smart City is a reference book for urban planners, architects, engineers, and property owners. Along with the variation in rainfall distribution, urbanization poses increased water demand and lessens infiltration and groundwater recharge. The modern "Water Smart City" concept arises with the interdisciplinary field to address design, construction, water management, and their interrelated responses towards them and the adjacent spaces. Harvesting rainwater in the urban system is the focus of this book to balance water scarcity and water abundance. Practical information has been provided for the rainwater collection and distribution, integrated with stormwater and greywater to reuse or manage aquifer recharge within residential and commercial properties. This book comprises fundamental and advanced topics presented in eight chapters: Introduction, Hydrological aspects, Rainwater harvesting system, Stormwater management, Greywater reuse, Groundwater recharge, Economics of rainwater harvesting system and Advanced technologies in the water smart city.

Chapter 1 *Introduction* provides an overview of urban water challenges, history of rainwater harvesting systems, brief conceptualization, implementation, and management. Then, the discussion is devoted to the potentiality of building a water smart city.

Chapter 2 *Hydrological aspects* describe the hydrologic process followed by the conventional hydrologic measurement. Hydrologic analyses emphasize establishing rainfall-runoff methods and extreme events analysis and then contribute to the hydrologic design.

Chapter 3 *Rainwater harvesting system* focuses on the component-wise conceptualization, design, implementation, operation, and maintenance to ensure harvested rainwater quantity and quality.

Chapter 4 *Stormwater management* is presented to facilitate the built environment in association with rainwater harvesting. Thus, technical interactions among the ecology, biodiversity, bio-inspiration, architecture, landscape and water values planning, natives' well-being, and socio-economic aspects are described in this chapter.

Chapter 5 *Greywater reuse* covers "greywater footprint", existing codes, water treatment, general consideration on design and installation, integrated uses with either rainwater or/and stormwater, economic assessment, and management strategies.

Chapter 6 *Groundwater recharge* highlighted managed aquifer recharge, relevant recharge technologies to ensure quantity and quality, and describes worldwide regulations and guidelines.

Chapter 7 *Economics of rainwater harvesting system* has described a realistic assessment of the life cycle cost and the benefit-cost ratio of the desired plan.

Chapter 8 *Advanced technologies in the water smart city* have started from the hydrologic process, designing rainwater harvesting systems, automated rainwater treatment, estimating stormwater, managing greywater, or/and managing aquifer recharge. This chapter also included socio-economic models and advanced tools for planners, architects, and engineers.

This book attempted to integrate rainwater harvesting into three significant fields, i.e., stormwater management, greywater reuse, and managed aquifer recharge. Many courses supplement this book at the undergraduate or graduate levels, with titles "harvesting urban rainwater", "smart rainwater management", "surface runoff harvesting", "managed aquifers in the water smart city", "smart water", etc.

This book has prepared in the S.I. unit and provided relevant example problems at the end of each chapter to illustrate the principles of analysis and the design procedure. If theoretical developments seem too extensive, an example problem has been included in the text to promote more accessible designing practices. A list of references has been included at the end of each chapter for supplementary reading. The book is profoundly illustrated with sketches.

This book's development is experience sharing based as a user and then the designer/engineer of the rainwater harvesting system. The indigenous rainwater harvesting practiced by my grandparents, who lived close to the Bay of Bengal, was the only fresh, rather non-saline water to drink. Since childhood, I was convinced rainwater could supplement only coastal areas' potable water demands until appointed to the Chittagong University of Engineering and Technology as a lecturer. Following my stay in the hilly parts of Chittagong since 2005, different research groups conducted research and work for a solution on the water abundance and water scarcity for the urban community. Learning from the discussion, knowledge sharing, field study, international networks, and research forum, I have started intensively working on *rainwater harvesting as a possible solution.* For the first time, WaterAid Bangladesh started the *Rain Day* celebration in Bangladesh. Being the regional keynote speaker in 2013 titled *Let's start rainwater harvesting* has become a part of this journey to date.

On the other hand, as recognition for research in engineering, one of the research works titled "Potentiality of Rainwater Harvesting for an Urban Community in Bangladesh" was awarded the prestigious Bangladesh University Grants Commission Award in 2015. Then, a more extended period has been spent learning and compiling a broader range of urban rainwater harvesting towards consistent and complete work. Some texts were taught in hydrology courses at the Chittagong University of Engineering and Technology (CUET). Based on classroom experiences, the delivery style has attempted to be interactive in this book.

I acknowledge my parents', Prof. H. Harun Ar Rashid and Rokeya Shamsun Nahar, endless support to reach here. I am grateful for the encouragement and moral supports of my husband, Prof. Dr. G. M. Sadiqul Islam, Department of Civil Engineering, CUET. Special thanks to the colleagues and students at the CUET, Bangladesh. Without the reviewers' comments and suggestions, this book would never be in shape—their essential guidance is highly appreciated. Appreciation goes to Eng. Tanvir Araf and Eng. Md. Redwoan Toukir for their valuable contribution to this book.

This book is a journey notebook on the systematic application and development of rainwater harvesting towards a water smart city. For the natives' well-being, any suggestions for improving the book are always welcome and will be incorporated in the next edition.

Chittagong, Bangladesh Aysha Akter

Contents

Abbreviation

ABC	Active Beautiful Clean
ABM	Agent-based modeling
ABS	Acrylonitrile Butadiene Styrene
AHP	Analytic Hierarchy Process
ANNs	Artificial Neural Networks
APHA	American Public Health Association
ARI	Average Recurrence Interval
ASR	Aquifer Storage and Recovery
ASTM	American Society for Testing and Materials
ASTR	Aquifer Storage, Transport, and Recovery
BASIX	Building And Sustainability Index
BIM	Building Information Modeling
BMPs	Best Management Practices
BMRC	Bio-Matter Resequencing Converter
BOD	Biochemical Oxygen Demand
CAD	Computer-Aided Design
CGP	Concrete Grid Pavers
CSOs	Combined Sewer Overflows
DAD	Depth-Area-Duration
DEMs	Digital Elevation Models
DN	Nominal Diameter
DOQQs	Digital Orthophoto Quarter Quadrangles
EV	Extreme Value
FAR	Floor Area Ratio
FRP	Fiber-Reinforced Plastic
FWS	Free Water Surface
GAC	Granular Activated Carbon
GEV	Generalized Extreme Value
GIS	Geographic Information System
GSI	Green Stormwater Infrastructure
GWF	Greywater Footprint

HVAC	Heating, Ventilation, and Air Conditioning
ICC	International Code Council
IDF	Intensity-Duration-Frequency
IDW	Inverse Distance Weighting
IPCC	Intergovernmental Panel on Climate Change
IRR	Internal Rate of Return
LCA	Life Cycle Assessment
LCC	Life Cycle Cost
LID	Low Impact Development
LiDAR	Light Detection and Ranging
LIUDD	Low Impact Urban Design and Development
LULC	Land Use Land Cover
MAR	Managed Aquifer Recharge
MARR	Minimum Acceptable Rate of Return
MBR	Membrane Bioreactor
MCA	Multi-Criteria Analysis
MCDA	Multi-Criteria Decision Analysis
MF	Micro Filtration
MGC	Maximum Ground Coverage
MOS	Mandatory Open Space
MUPA	Mandatory Unpaved Area
NF	Nano Filtration
NOAA	National Oceanic and Atmospheric Administration
NPV	Net Present Value
NRCS	Natural Resources Conservation Service
NURP	Nationwide Urban Runoff Program
NWS	National Weather Service
PB	Payback Period
PICBP	Permeable Interlocking Clay Brick Pavers
PICP	Permeable Interlocking Concrete Pavers
PMP	Probable Maximum Precipitation
POE	Point-of-Entry
POU	Point-of-Use
PTRG	Plastic Turf Reinforcing Grids
PVC	Polyvinyl chloride
RADAR	RAdio Detection And Ranging
RO	Reverse Osmosis
RSF	Rapid Sand Filter
RVE	Representative Volume Element
SCS	Soil Conservation Service
SOCs	Synthetic Organic Chemicals
SPSS	Statistical Package Social Science
SSF	Slow Sand Filter
SuDS	Sustainable Drainage Systems
TED	Top of Extended Detention

TRRL	Transport and Road Research Laboratory
UAV	Unmanned Aerial Vehicle
UF	Ultra Filtration
UNEP	United Nations Environment Program
US EPA	U.S. Environmental Protection Agency
UV	Ultra Violet
VOCs	Volatile Organic Chemicals
WMO	World Meteorological Organization
WSUD	Water Sensitive Urban Design
YAS	Yield After Spillage
YBS	Yield Before Spillage

Chapter 1
Introduction

1.1 General

Urbanization poses increased water demand due to rapid population growth, and also increased impervious lands exaggerate surface runoff and lessen groundwater resources. On the other hand, climate change significantly influences the quantity and quality of rainfall. Worldwide urban water management faces three-dimensional challenges, i.e., potable water shortage, urban floods, or waterlogging and depleting groundwater. A properly planned, designed, and constructed rainwater harvesting system has been practicing as a simple approach within the city context to improve the community's wellbeing. Despite the advantages of harvesting rainwater as a sustainable development to mitigate urbanization-related issues, there is a knowledge gap within the design and management features to deliver significant outcomes. This chapter provides an overview of urban hydrology, challenges on urban water management, the glorious history of rainwater harvesting systems, briefly implementation and direction, and their potentiality towards a water smart city. Guidance on the site selection, maximization of harvesting quantity, and combination with stormwater and greywater are summarized to provide insights for overcoming rainwater-related concerns. Worldwide available codes and legislation for rainwater harvesting have also included, thus, the inevitable noteworthy role of the decision support system supposed to be encouraged.

1.2 Urban Hydrology

Worldwide rapid urbanization along with climate changes poses a threat to the natural hydrological cycle. Globally, 55% of the population lived in urban areas than the rural areas in 2018, whereas this amount was 30% in 1950 and projected to be 68% in 2050 (UN 2018). The urbanization growth rate persists, faster growth is expected in Africa and Asia, and the projected urban residences are 56 and 64% in 2050, respectively,

© The Author(s), under exclusive license to Springer Nature Switzerland AG 2022

A. Akter, *Rainwater Harvesting—Building a Water Smart City*, Springer Water,
https://doi.org/10.1007/978-3-030-94643-2_1

compared to 43 and 50% in 2018 (UN 2018). Thus, faster urbanization would significantly influence all the three pillars of sustainable development, i.e., economic, social, and environmental. Urbanization results in rapid water table depletion to meet the increased demand, runoff due to incremental paved areas comprised of buildings, car parks, business complexes, and landfill dumping towards water bodies. In the recent urban century, cities face multiple challenges with urban hydrology.

'Urban hydrology' describes land hydrology, which investigates the hydrological cycle, water regime, and quality within the urbanized territory. *Urban hydrology* is a science linking among other sciences dealing with ecological problems, 3R's, i.e., 'Reduce', 'Reuse' and 'Recycle' for earth's water resources. Current and future professionals involved in local governance and urban planning, development, and implementation must cope with the rapid changes in the economy, society, urbanization, and climate. Thus, they should be aware of sustainability concepts while dealing with urban hydrology for planning, designing, and procuring buildings to ensure climate change mitigation and adaptation.

1.3 Water Challenges

Growing urbanization alters the hydrologic cycle by over-extracting groundwater resources, insufficient drainage facilities, and changing rainfall trends.

1.3.1 Ground Water Shortage

Since time immemorial, groundwater has become an essential part of the world due to the better quality than surface water, especially in the dry part. The stakeholders often affect widely accessible groundwater sources, whereas the prominent actors are farmers, industries, exploiters of quarries, and mining sectors. Similarly, citywide local water supply and sewerage service providers, local city development authority, supervisors of public or private works related to hydraulic engineering, and underground stakeholders include urban zones, traffic, and ground storage.

Usually, the water supply of Asia's high urban growth countries depends partially or entirely on groundwater abstraction (Fig. 1.1). There is a relationship between rainfall—groundwater recharge, and urbanization. Due to the absence of knowledge or to cope with urbanization, these actors often disrupt terrain conditions, aquifer structure, slow down the recharging process and pollute the groundwater storage with hazardous materials. The statistics on 45 countries worldwide showed that the domestic use based on groundwater abstraction ranges from 50 to 100% (Appendix A) (Margat and Gun 2013). Thus, there is an urgent need to adopt the water smart city concept from now on.

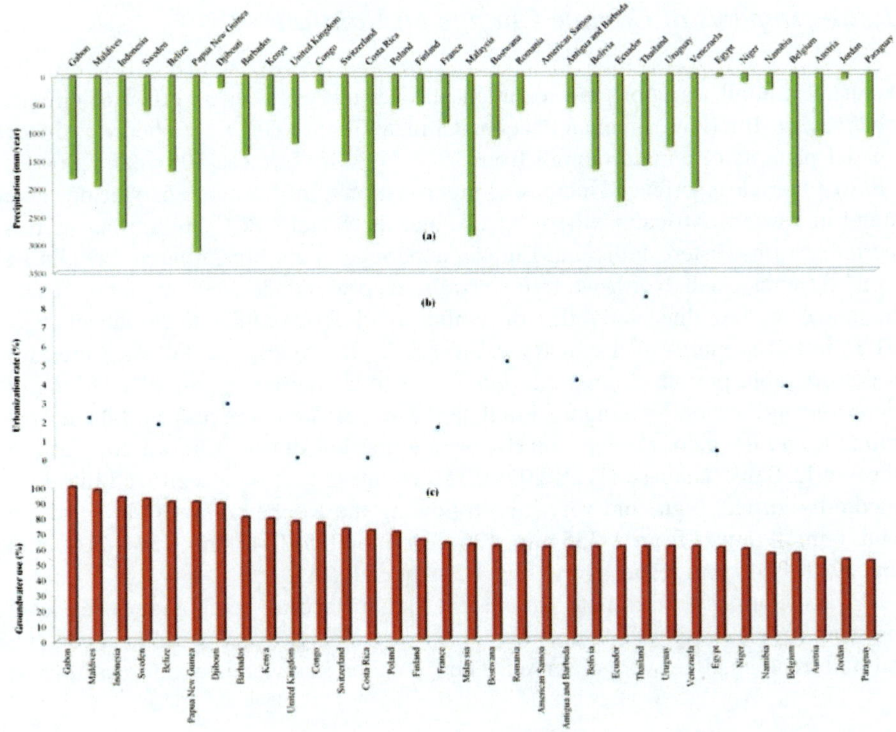

Fig. 1.1 Countrywise relation among **a** rainfall, **b** percentage of groundwater use, and **c** percentage of urbanization

1.3.2 Urban Storm Waterlogging or Urban Flooding

Due to changes in Land Use/Land Cover (LULC) and the progression of urbanization, the excess rainfall faces limited access to be absorbed by the adjacent lands and disposed of through the dedicated drains. Thus, in proportion to the urban growth rate, increased urban flood or storm water logging poses a great concern to the city areas during monsoon. Urban flood or urban storm waterlogging was identified worldwide as an obvious occurrence during monsoon, for instance, in cities within the US, Canada, Europe, Australia, Philippines, Sri Lanka, Japan, and China. Based on the topography, disposal of the surface runoff varies from the 'urban storm waterlogging' or 'urban flooding'. Experiences around the world showed that detailed knowledge of LULC and the hydrological cycle could minimize waterlogging. Rainwater harvesting showed innovation in reducing the quantity of urban storm waterlogging or urban flooding between 20 and 50% (Döll et al. 2012) and water pollution.

1.3.3 Impacts of Climate Change on Rainfall

Unusual rainfall variations are recorded all over the world as an effect of climate change. The Intergovernmental Panel on Climate Change (IPCC) studied worldwide spatial patterns of annual rainfall from 1901 to 2005. In these 105 years, most of North America experienced increasing precipitation trends compared to the negative trend in western Africa, southern Africa, and the Sahel (IPCC 2008). During this period, North-western India and Eurasia experience both escalating and declining rainfall trends. Also, North-western Australia reported moderate to higher increases in annual rainfall; thus, variability of rainfall trend observed due to climate change. The rainfall scenario of a country expresses by the yearly rainfall measures for a considerable period. *Annual rainfall* is the total measured rainfall in terms of depth during 12 months; annual rainfall has been classified as excess rainfall, normal rainfall, and deficient rainfall. Yearly average rainfall during 1960 to 2017 as per the World Bank database (WDI 2018), 156 countries can be categorized into low, medium, normal, high, and very high following the Köppen climate classification with rainfall ranges from 0–435 mm, 436–934 mm, 935–1543 mm, 1544–2200 mm and 2201–3240 mm respectively (Fig. 1.2; Appendix A).

In association with climate parameters viz. surface air temperature, sea level pressure, free atmospheric temperature, tropopause height, and ocean heat content, rainfall trend changes were observed. Thus, detailed knowledge and forecasting on

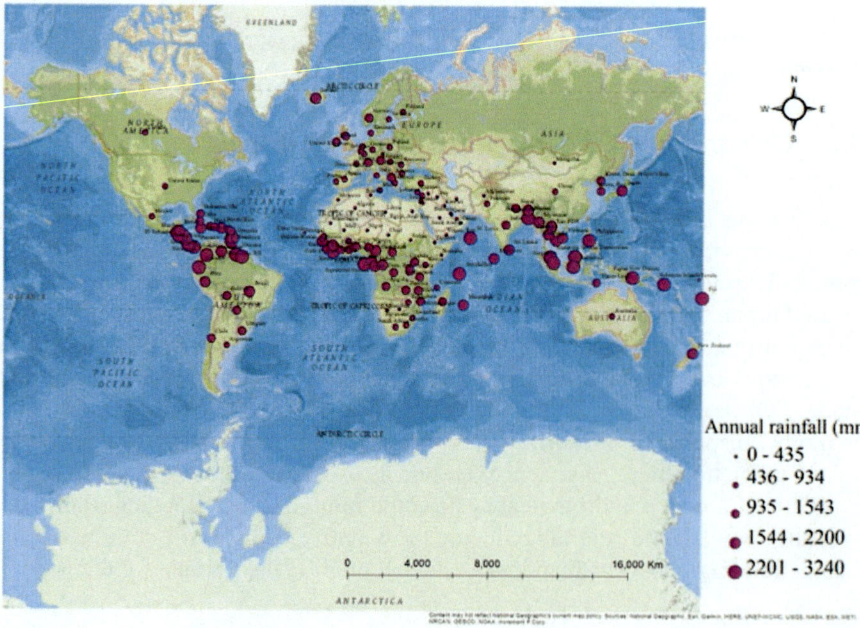

Fig. 1.2 Annual rainfall around the world (based on the available station data for 156 countries)

duration, intensity, and frequency of rainfall influence the planning, designing, and construction of a rainwater harvesting system.

1.4 Water Smart City

Water smart city in an integrated approach comprises of urban developments and urban water management. This approach has been researching for the last two decades; the three required pillars (Wong and Brown 2009) are:

- Cities act as water supply catchments: Cities comprised diverse water resources through an integrated centralized and decentralized infrastructure at different scales. Thus, cities need easy access to water at least environmental, social and economic costs.
- Cities provide ecosystem services and increase liveability: Once the combination of urban design and nature-based technologies is ensured, sustainable water management would be confirmed for the wellbeing of the city dwellers.
- Cities comprise water-smart communities and institutions: stakeholders are aware of the water-smart city concept. To converse a city's natural resources, reflective practitioners through industrial and professional capacity build-up and government policies could facilitate ecologically sustainable lifestyle in the city.

Harvesting rainwater becomes an essential part of building a water smart city and promoting this renewable resource country-wise different codes and legislations are already practicing. Also, to conserve water resources, rainwater harvesting-related financial incentives are announced.

The benefits of adopting a water smart city utilizing an integrated water cycle management approach include:

- Minimization of potable water consumption using rainwater harvesting and wastewater reuse where appropriate.
- The protection of water quality in the waterways by improving the quality of water draining from developments.
- The reduction in runoff and peak flows through on-site stormwater detention.
- Blue-green infrastructure in the streetscape can have a cooling effect and buffer the urban heat island effect.

1.5 Rainwater Harvesting System

A rainwater harvesting system is the technology or method of accumulation, conveyance, and water storage from rainfall in an area rather than runoff. The available methods are for collecting as well as rainwater storing from rooftop, land surface or catchments/watersheds through:

- Capturing runoff from the rooftop, i.e., rooftop rainwater harvesting
- Storing runoff from the local catchment, i.e., the land harvesting
- Arresting seasonal floodwaters from adjacent streams
- Conserving water through catchment/watershed management.

In urban areas, both roof and land-based rainwater harvesting systems have been practicing. The *rooftop* includes public, private, commercial, industrial buildings, and open spaces contain pavements, lawns, and gardens. The essential components of this system are catchment, conveyance network, rainwater storage tank, pump, and harvested water distribution. To improve the required water quality of the utilized rainwater, the inclusion of treatment technologies is practicing before or /and during storage. In addition to these, for an emergency, a makeup water supply and overflow options are incorporated to meet the low rain period and excess rain period, respectively. Thus, the system's efficiency depends on rainfall, runoff collection, storing water, and distribution of the harvested water. In cities, harvested rainwater extends up to potable after the various range of harvested water treatment. Already for non-potable uses both indoor (i.e., toilets flushing, laundry) and outdoor, i.e., watering the garden and irrigating landscapes, washing cars, sidewalks, roads, etc. in water fountains and water features for recreational purpose, fire suppression viz. fire trucks and fire hydrants, water cooling towers and so. In a swimming pool, the harvested water uses after proper treatments.

With approved treatment and measures, harvested water would be an excellent means of potable water and culinary. Thus, rainwater harvesting is getting attraction in cities due to:

- Precipitation, i.e., rain, hail, sleet, and snowfall, are naturally distilled through evaporation before cloud formation;
- Absence of calcium carbonate, magnesium, potassium, and sodium salts, rainwater is the most desirable water source on our planet. Thus, this is the cost-effective, simple technology with easy operation and maintenance;
- Reduction of water bills and energies while using harvested water;
- Reduces groundwater demands and can contribute to recharging groundwater;
- Rainwater believes in the natural fertilizer as this contains sulfur, nitrogen, also microorganisms, and mineral nutrients collected from dust in the air; and
- Minimize urban storm waterlogging or floods in the city.

In contrast to the benefits mentioned above, the urban rainwater harvesting system often becomes a burden to the owners due to:

- Unpredictable rainfall while the designer failed to work on future rainfall trend prediction.
- The high capital cost is due to the rainwater harvesting system installment and design storage size and technology.
- Proper maintenances are needed both in the water collection and storage stages.

1.6 History of Rainwater Harvesting

The glorious history of *rainwater harvesting* for more than 10,000 years worldwide focused on storage, and gradually other components got modifications. From ancient ages to modern life, archaeologists have discovered several sophisticated collections, conveyance, and storage systems for rainwater harvesting systems worldwide as the main or supplement water source. The modes of harvesting system are based on circumstances, resource availability, and trends or culture of a particular place.

1.6.1 Ancient History

The history of the *rainwater harvesting system* observes in the Middle East and Asia (Gould and Nissen-Petersen 1999). During the Bronze Age, rainwater harvesting was practiced and comprised various techniques for collection to supply; many of those are practicing in modern ages. Through the Neolithic Age, to store rainwater, waterproof lime plaster cisterns were traced in the village of the Levant, located in Southwest Asia (Miller 1980; Mays et al. 2013). In the Mediterranean region, rainwater harvesting was used at the Palace of Knossos during 1700 BC (Mays et al. 2013). In Europe, through the Minoan period (2,600 BC – 1,100 BC), the existence of large cisterns was found in the Greek island of Crete (Mays et al. 2013) (Fig. 1.3). The earliest masonry-lined cisterns were discovered for rainwater storage in ancient Greece during Minoan civilization in the first half of the 2nd millennium BC (Angelakis and Spyridakis 2010).

Fig. 1.3 Minoan cistern at the housing complex in the vicinity of the village Chamaizi, Sitia (Angelakis and Spyridakis 2010)

Within late 4000 BC, dry-land farming cisterns offer potential water management (Mays et al. 2013). Jerusalem and Israel have glorious cistern heritage, a large cistern of around 2500 BC found in the Biblical city of Ai (Khirbet et-Tell). Solid rock and significant stone linings were used in this cistern and sealed with clay to minimize leakage (Mays et al. 2013). Most water harvesting structures viz. reservoirs, tanks, and canals were constructed for irrigation and supported household activities. Farming communities, especially in Pakistan, India, and Bangladesh, have been harvested rainwater for more than centuries. In the Chola period (1003 and 1010 AD), the Shivaganga tank and Vīrānam tank serve multi-purposes. Almost 2,000 years back in Thailand, roof catchment rainwater collection via simple gutters into traditional jars and pots traces.

In ancient times, earthen dams and cisterns usually served as storage with simple gutters to fill jars and pots by the rainwater. In Thailand, the use of gutters can be found back almost 2000 years. Ancient rainwater cisterns are still available on the islands of Capri and Malta. Ancient Africans of the desert zone so far stored rainwater for drinking purposes. Romans discover rainwater harvesting as air conditioners, evaporation of the rainwater (from pools, reservoirs, etc.) to offer an air-conditioning feeling. This technology extended to connect underground cisterns to above-ground pools for water filtration. Once these pools overflowed, the cisterns were filled with cleaner water. Later, this design inspired modern-day *rain barrels*. On the other hand, Roman's shallow pool motivated the roof washer or *first flush system*.

1.6.2 Modern Time

In the modern era, the use and application of rainwater harvesting have throughout the world. In 2009, the United Nations Environment Program (UNEP) highlighted the increasing trend of rainwater collection techniques and was recognized as a possible means for water-scarce and water-rich countries. British government's code in 2015 stated that the ideal designs had the potential to reduce up to 50% of demand on mains water supply. Thus, harvested rooftop rainwater can serve domestic (i.e., laundry and lavatory plumbing), livestock, gardening, and irrigation. Urbanized areas have also started adopting rainwater harvesting systems through stormwater reduction and groundwater level replenishment. After the proper treatment, cities like Texas, Ohio, Beijing, Singapore, and Mumbai have adopted rainwater harvesting. The most significant rooftop rainwater harvesting projects are in China (Gansu province) and semiarid northeast Brazil.

Rural areas traditionally practice rainwater harvesting based on local material as well as technology. Rainwater harvesting usually practices most west and south rural housing in New Zealand due to plentiful rainfall and is promoted by the existing council policies. Earthen rainwater ponds in the Irrawaddy Delta of Myanmar serve the drinking water to avoid saline water throughout the dry season. Many of these ponds aged a century or more and pose a great heritage within the society. Coastal areas and hilly parts in Bangladesh practice harvesting rainwater in a clay pot for the dry season. In India, to date, rainwater harvesting practicing to minimize groundwater table lowering (viz. Tamil Nadu, Andhra Pradesh, etc.), water supply (Mumbai, Maharashtra, Bangalore, Rajasthan, etc.), "*Rainwater harvesting theme park*" (Karnataka, the Thar Desert in Rajasthan, etc.). In Senegal and Guinea-Bissau, ethnic groups (i.e., Diola-people) use local and organic materials to practice homebrew rainwater harvesters. Around 40% rural area of Thailand is dependent on harvested rainwater due to the Thai government's water conservation initiation during the 1980–1990s. Thai private sectors continued to provide several million tanks for household uses to support these initiations, turning worldwide one of the most extensive self-sufficient water supplies. A similar initiation was in 2003, a program called "Programa Um Milhão de Cisternas" ("One Million Cisterns") in Brazil with rooftop rainwater harvesting systems for one million homes. To minimize groundwater depletion during 2001, Tamil Nadu in India has initiated compulsory rainwater harvesting for urban and rural households. There are many initiatives for rainwater harvesting in the rural areas of Karnataka, ranging from the construction of small check dams, farm ponds, bunds, and recharging of existing bore wells by channeling the rainwater during monsoon to slits cut into the casing of the bore well.

In the modern age, the adaption of rainwater harvesting remains a choice and a necessity. As the water demands vary from country to country, the laws and regulations are stipulated even within the country. Canadian law since the mid-2000s could increase the trend in rainwater harvesting technology. Thus, the law emphasized rainwater harvesting within all new construction in Bermuda residents. This practice is similar to the Virgin Islands in the United States (US) and Santa Fe, New Mexico. Texas offers a sales tax exemption on rainwater harvesting equipment. Until 2009, to secure surface runoff from the watershed in Colorado, water rights laws were restricted rainwater harvesting for the property owner. In 2007, a study in Douglas County, in the southern suburbs of Denver, identified loss in the hydrological cycle reduces up to 97% of the precipitation to reach the nearby stream. These findings influenced the Colorado legislature; later on, the amended law supported the property owners with a rooftop rainwater collection system and ten large-scale pilot studies. The *water for 2060 Act* was passed in 2012 to promote rainwater and greywater use, among other water-saving techniques in Oklahoma. Germany has promoted rainwater harvesting practices since the 1980s. In the United Kingdom (UK), traditional rainwater storing water barrels (or butts) for gardening motivated the revoked British government's code in 2015. This code encouraged large underground tanks to newly constructed residences for harvested water in non-potable use.

In South Asia, Urban Development Authority Sri Lankan promotes rainwater harvesting during 2007. Different states of India implemented rainwater harvesting as compulsory for any new residential community, for instance, Pune and Bangalore. Mumbai and Maharashtra are in the process of establishing mandatory rainwater harvesting for large societies. In 2020, Bangladesh included rainwater harvesting in the building code. Likewise, water research commissions around the world conduct researches on rainwater harvesting.

Researchers and practitioners are working on modern concepts and ideas. Dutch invention 'Groasis waterboxx' store rainwaters in a box contain plant or tree and provide enough water for its growth with harvested precipitation (Mays et al. 2013). Rainwater harvesting by rain saucer using solar power polyvinyl panels (Leung 2008) is research to harvest palatable water with limited filtration and disinfection processes. The eco classroom concept is an eco-friendly space to provide sufficient drinking water and solar energy for the resident. The eco classroom Benenden is considered the first building within the UK, mainly designed to harvest rainwater to supply potable drinking water using solar energy (Inhabitat 2014). Electricity generation by rainwater runoff is also attracting city dwellers. Nature-based technologies are designed to capture excess rainfall. These include bioswales, bio-retention cells, rain gardens, green roofs, infiltration trenches, continuous permeable pavement, rain barrels (or cisterns), rooftop disconnection, bench terraces, check dams, contour bunds, and contour ridges.

1.7 Advantages of the Rainwater Harvesting System

Rainwater harvesting has a broader range of worldwide acceptability among flat plain to hilly terrains, saline or coastal areas, islands, and desert. However, in general, the following advantages are for this system to build a water smart city:

Quantity:

- Supplement or improved water supply; and
- Ensures self-sufficiency to water supply.

Quality:

- Offers high-quality water, soft and low in minerals
- Improves groundwater quality through dilution when recharged to groundwater; and
- Reduces soil erosion in urban areas.

Management and economics:

- Rooftop rainwater harvesting is a cost-effective system
- Simple to construct, operate, and maintain; and
- Lowers pumping costs for groundwater withdrawal.

Within a city:

- Lessen the demand on municipal water systems
- Escape strict watering schedules
- Ensure nutrient-rich rainwater to the landscape than municipal water
- Reduces chemicals exposure; and
- Reduces fertilizer usage for gardening.

1.8 Selection of Rainwater Harvesting System

The selection of rainwater harvesting systems in a city depends on total precipitation, LULC, rainfall frequency and duration, dry periods, catchment efficiency, and user choice. Following considerations are required:

- Purpose of rainwater harvesting system: water supply, urban flood control, and recharge to groundwater;
- Assessment of rainwater as a potential option- simple mass balance approaches or 'rule-of-thumb' based on annual precipitation usually used for performance study;
- The temporal and spatial variation of rainfall along with the other surface or groundwater availability;
- Design period;
- The designed storage capacity of rainwater;
- Development of a reference map or computer tools using behavioural (stochastic) models for sizing of storage volume for rainwater harvesting systems;
- Types of water storage structures and their selection;
- Distribution techniques;
- The economic status of the consumers;
- Operational and maintenance strategies; and
- Hydrological impact of rainwater harvesting systems.

1.9 Maximizing Harvested Water Efficiency

Integrated rainwater harvesting systems need to be designed considering the users' demand, water availability, economic status, geology, topography, and urban hydrology. Rainwater harvesting can be practiced on built-up areas only, non-built-up areas only, or/and an integrated approach in rainwater harvesting structures and land use. A combination of rainwater harvesting structures would include *recharge trenches*, sump, percolation pond, etc.

- Better yield outcomes ensure through maximize the roof area, accumulating rainwater to the harvesting system. The minimum roof area should be taken as the regional standard or the greatest of 50% of the roof area or 100 m^2 (Rainwater Harvesting Association of Australia and urban water cycle solution), details in catchment section, Chap. 3. The selection of roofs/catchment areas plays a vital role in maximizing their Rainwater Harvesting potential. A broad range of runoff coefficients is involved in runoff calculations due to the interactions of various factors, i.e., climatic (size and intensity of the rain event, antecedent moisture, prevailing winds) and architectural (slope, roof material, surface depressions, leaks/infiltration, roughness). Thus, for the conveyance, a properly installed roof guttering system on a standard roof would connect approximately 100 m^2 of roof catchment to the rainwater tank via two downpipes. An option to connect more roof areas might involve using a charged downpipe connection to the rainwater tank.

- Conveyance connects to all taps for outdoor (gardening, car wash) and indoor (toilets, washing machine, and whole house) usage to maximize efficiency. A connection to the main water supply acts for both mains/supplement water through a bypass in dry weather (in the absence of rainwater within the tank).
- The use of short suction lines might reduce air entrapment issues. However, this could lose the prime and effective operations of the pump. The pump position should allow minimum elbows in the suction. The presence of elbows in the suction line causes noises and weak pump performances.
- A flexible suction line is recommended due to the absence of elbows and ensures easy conveyance from the rainwater tank and the pump. Flexible suction pipework follows the setup of loops and should direct downwards. This would prevent trapped air in the high point and reduce issues with pump performances.

Storing rainwater might prevent aquifer from recharging as well as proper planning can contribute to depleting groundwater trend. Conjunctive use of groundwater and rainwater could work on aquifer replenishment. Then, conjunctive use of surface water and rainwater might offer a potable water supply for the city dwellers with reduced costs. Similarly, conjunctive use of rainwater and greywater could save the usage of palatable water for car washing, toilet flushing, gardening, and so on. Most of these approaches have already been practicing throughout the world.

1.10 Rainwater Harvesting System-Related Codes and Legislation

Harvesting rainwater becomes an essential part of the water smart city, and country-wise different codes and legislations are already practicing to promote this renewable resource (Table 1.1). Also, to conserve water resources, rainwater harvesting-related financial incentives are announced. Among the published codes or standards, rainwater harvesting guidance for consumers is in three modes—starting within plumbing codes to permit harvested water for secondary purposes. Rainwater harvesting in building code to minimize storm waterlogging and latest codes include conjunctive use with greywaters.

Australian Government has taken steps to ensure energy-efficient and water-efficient designs and products for the newly constructed houses. For example, Victoria imposes a 5-star energy efficiency rating for the building fabric, water-efficient taps, fittings, a rainwater tank for toilet flushing, or a solar hot water system. South Australia requires an indoor rainwater tank. In Syndey and New South Wales, the BASIX (Building And Sustainability Index) building regulations undertakes a 40% reduction in mains water usage. The BASIX target for water conservation includes (i) Showerheads, tap fittings, and toilets with at least a 3A rating; and (ii) A rainwater tank or alternative water supply provision.

Table 1.1 Rainwater harvesting related codes around the world

Country	Code	Contribution to the practice
Australia	BCA 2006	Rain tank design
Bangladesh	BNBC 2020	Rooftop rainwater harvesting for potable water and groundwater recharge in the city
Canada	CMHC 2012	Guidelines for residential rainwater harvesting
India	Indian Standard 2008	Rooftop rainwater harvesting for groundwater replenishment
UK	BSI 2018	Rainwater control systems
US	New Mexico, Santa Fe County 2003	Rainwater catchment for both commercial and residential developments
	Washington state legislation 2003	Exemption of income tax for rainwater harvesting
	Oregon (Building codes division 2008)	Construction of rainwater harvesting
	North Carolina 2009	Harvested rainwater for toilet flushing and irrigation
	State of Colorado 2009	Promote water conservation through rooftop rainwater harvesting
	State of Rhode Island 2012	Exemption of income tax for rainwater harvesting
	California plumbing code 2013	Rainwater and greywater as alternative sources

On the other hand, a 3,000-liter rainwater tank in the Pimpama Coomera Master Plan area of Gold Coast is connected to the recycled water system. The state of Queens in Australia offers a rebate of US $1200 to purchase and install home rainwater storage. In Germany, property owners collect rain taxes based on the impervious surface and generate direct storm runoff. So, the increased rainwater conservation offers less storm runoff, and this would require lesser storm sewers. Thus, converting impervious pavement/roof into a porous surface would offer rain tax. In Arizona, the government provides a tax exemption of 25% of the cost of a water conservation system, including residential greywater and rainwater.

Similarly, New Mexico and Texas are practicing tax exemption on the assessed value of the respected property with a modified water conservation system. Madhya Pradesh in India offers a rebate of 6% on property tax as an incentive for the rainwater harvesting systems. In most cases, the building having a specific floor size accounts for rainwater harvesting suitability. For instance, in Bangalore in India, rainwater harvesting is compulsory for the property owner or the dwellers on a floor size 223 m^2 and a newly constructed building of 111 m^2 and more. Similarly, the Bangladesh building code permits the floor size should be more than 300 m^2 for mandatory rainwater harvesting.

References

Angelakis AN, Spyridakis DS (2010) A brief history of water supply and wastewater management in ancient Greece. Water Sci Technol Water Supply 10(4):618–628

BCA (2006) Building code of Australia. https://www.builderassist.com.au

BNBC (2020) Bangladesh national building code. Ministry of housing and public works

BSI (2018) On-site non-potable water systems. Systems for the use of rainwater. BS EN 16941-1:2018. 978 0 580 89028 4

Building codes division (2008) Approval of rainwater harvesting systems as a statewide alternative method of providing water for non-potable uses. OPSC 08-03. State of Oregon

California plumbing code (2013) Alternate water sources for nonpotable applications, 6 and 7, pp 307–332. http://www.arcsa-edu.org

CMHC (2012) Guidelines for residential rainwater harvesting systems handbook. Canada Mortgage and Housing Corporation (CMHC)

Döll P et al (2012) Impact of water withdrawals from groundwater and surface water on continental water storage variations. J Geodyn 59–60:143–156. https://doi.org/10.1016/j.jog.2011.05.001

Gould J, Nissen-Petersen E (1999) Rainwater catchment systems for domestic supply: design. Intermediate Technology Publications Ltd., Construction and Implementation

IPCC (2008) In: Eggleston HS., Miwa K, Srivastava N, Tanabe K (eds) 2006 IPCC guidelines for national greenhouse gas inventories—a primer, prepared by the national greenhouse gas inventories programme. IGES, Japan.

Indian Standard (2008) Roof top rainwater harvesting—guidelines, WRD 3: ground water and related investigations. IS 15797: 2008

Inhabitat (2014) Solar-powered blue forest eco classroom harvests drinking water from its green roof. https://inhabitat.com

Leung J (2008) Rainwater harvesting 101. Council on the environment of NYC (CENYC). cupdf.com/document

Margat J, Van der Gun J (2013) Groundwater around the world: a geographic synopsis. CRC Press

Mays L et al (2013) History of water cisterns: legacies and lessons. Water 5(4):1916–1940, Multidisciplinary Digital Publishing Institute. https://doi.org/10.3390/w5041916

Miller R (1980) Water use in Syria and Palestine from the Neolithic to the Bronze Age. World Archaeol. 11(3):331–341, Routledge. https://doi.org/10.1080/00438243.1980.9979771

North Carolina (2009) General assembly of north Carolina. Session law 2009-243; house bill 749.

Santa Fe County (2003) Santa Fe County land development code. Ordinance No. 2003-6. https://www.santafenewmexican.com

State of Colorado (2009) Concerning an authorization of pilot projects for the beneficial use of captured precipitation in new real estate developments, and making an appropriation in connection therewith.

State of Rhode Island (2012) Relating to taxation—personal income tax. H 7070. LC00160. http://webserver.rilin.state.ri.us

UN (2018) World urbanization prospects: the 2018 revision, highlights (ST/ESA/SER.A/352)

Washington state legislation (2003) Stormwater control facilities—rates and charges—limitations—use. Washington state. https://apps.leg.wa.gov/RCW

WDI (2018) WDI database: world development indicators. The World Bank, 1818 H Street NW, Washington, D.C. 20433-USA

Wong THF, Brown RR (2009) The water sensitive city: principles for practice. Water Sci Technol 60(3):673–682. https://doi.org/10.2166/wst.2009.436

Chapter 2
Hydrological Aspects

2.1 Introduction

The young science 'hydrology' reached mathematical analysis for many hydrological principles and involvements of sophisticated instruments and computer techniques. Rainwater harvesting is solely dependent on the 'hydrological cycle.' Adequate knowledge is required to address issues related to urban hydrology for planning, designing, and implementation, and afterward. Hydrological cycle alterations on the increased impervious lands cause increased surface runoff. Initially, with a detailed hydrological cycle description, water-related challenges toward a water smart city are listed. This chapter describes the conventional measurement and computation of precipitation, infiltration, surface runoff, and groundwater movement. During hydrological cycle alterations, change in water quality is also noticeable. Finally, this chapter contains relevant workout example problems. The prime element for rainwater harvesting is achieved through identifying 'design rainfall/ precipitation. A detail on precipitation formation, distribution, frequency, and mapping has illustrated in this chapter. From the 'rainfall-runoff' and 'rainfall intensity-duration-frequency' relationships, along with land use land cover, surface runoff was obtained for stormwater management. Knowledgeon infiltration and groundwater movement requires designing artificial aquifer recharge, i.e., managed aquifer recharge. This chapter suggested the necessities for advancing the described measurement and computation processes in the hydrological aspects.

2.2 Hydrological Cycle

'Hydrology' is a field of science interrelated with other branches of science to deal with water considering the properties, distribution, and circulation. Professor Ven Te Chow had divided the development of hydrology into eight phases, and these

© The Author(s), under exclusive license to Springer Nature Switzerland AG 2022

A. Akter, *Rainwater Harvesting—Building a Water Smart City*, Springer Water,
https://doi.org/10.1007/978-3-030-94643-2_2

are speculation (up to 1400 AD), observation (1400–1600), measurement (1600–1700), experiment (1700–1800), modernization (1800–1900), empiricism (1900–1930), rationalization (1930–1950) and theorization (1950–till date). The *hydrological cycle* exhibits water movement from the sea, passed over lands, and discharge to the sea. Thus, the hydrological cycle, as a case of circular infinity, progresses with the three phases of the earth system, and these include Hydrosphere (water bodies over the earth surface), Atmosphere (gaseous envelope around the planet), and Lithosphere (soil layers and rocks underneath the hydrosphere). The hydrological cycle globally concerns the water movement processes on the atmosphere, land surface, and sub-surface. A hydrological cycle study could start from any of the processes, i.e., precipitation, evaporation, transpiration, condensation, interception, infiltration, percolation, runoff, and storage (Fig. 2.1).

- **Atmosphere zone**: *Precipitation* deals with all the moisture emitting from the atmosphere and return to the land.
- **Land surface zone**: receiving precipitation from the atmosphere and generates outflow as (i) evaporation returns to the atmosphere, (ii) infiltration through the sub-surface zone, and (iii) surface runoff. In the *Evaporation* process, water vapour passes from water bodies or landmasses and finally diffuses into the atmosphere. Evaporation takes place from land, lakes, streams, oceans, and during precipitation. *Transpiration* comprises a water passage from liquid to vapour state through

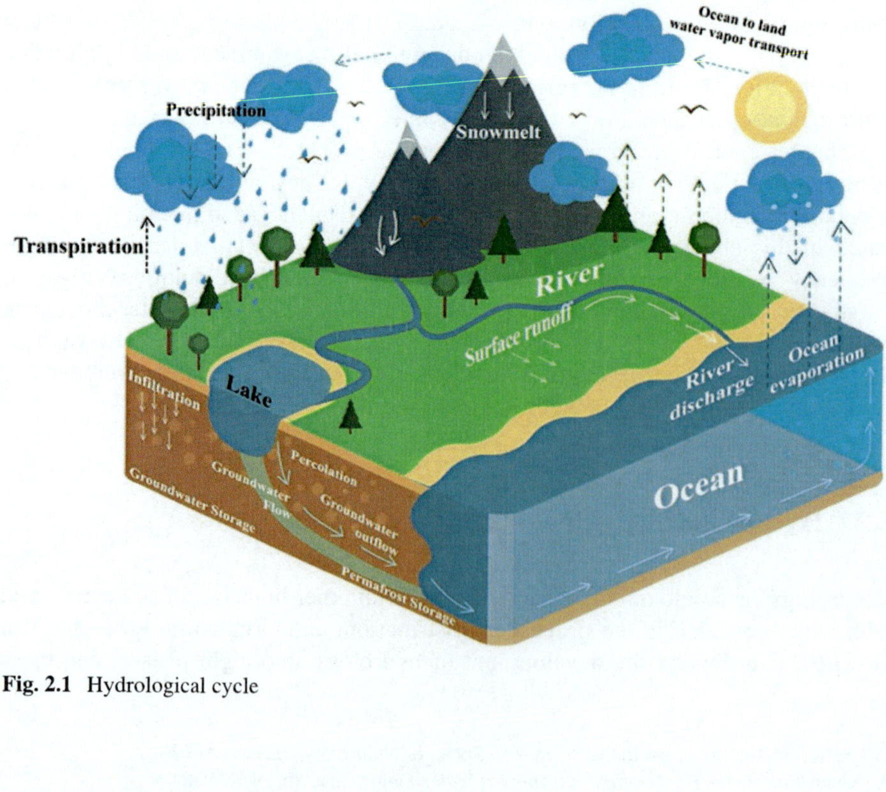

Fig. 2.1 Hydrological cycle

plant metabolism. Evaporation and transpiration often describe together as 'Evapotranspiration'. *Vapour diffusion* includes a portion of water retained by the soil, passes as vapour towards the ground surface. *Interception* usually captures a part of precipitation and influenced by the vegetation and evaporates moisture. *Surface runoff* deals with the precipitated water flow in a stream after meeting all the requirements of surface and sub-surface. *Depression storage* or surface retention enables precipitation over a catchment stored in ditches, ponds, sinks, lakes, etc. Surface detention is the temporary storage in the river channel.
- **Sub-surface zone**: *Infiltration* is the penetration of a portion of precipitation into the ground and afterward flows downwards. Interflow during infiltration, water starts moving laterally towards a stream and appears on the surface is known as *interflow*. Interflow exists above the groundwater table, while the flow velocity is lower than the surface flow. Then, interflow is known as sub-surface storm flow, sub-surface runoff, storm seepage, and secondary base flow. The infiltrated water might reach the saturated zone of water below ground and then get stored among the pores and voids between particles. Water movement underneath the ground surface is known as *groundwater flow*.

2.3 Urban Hydrology

A water smart city is supposed to overcome the adverse impact on the interactions between land and water continues as urbanization proceeds. *Urban hydrology* describes as the interdisciplinary science of water and human interference with natural processes (Leopold 1968; Johnson and James 1971; Niemczynowicz 1999). Typical urban hydrologic cycle faces hydrologic changes in each zone associated with increased imperviousness:

- **Atmosphere zone**: along with precipitation generation and pollutants deposition progresses on the land surface zone.
- **Land surface zone**: evaporation, infiltration, and surface runoff occurs in this zone. Additionally, pollutant loadings intrude with the surface runoff.
- **Subsurface zone**: infiltration occurs in this zone, and a portion of inflow transfers to the conveyance zone.
- **Conveyance zone**: comprises link of hydraulic elements viz. drains or channels, pipes, pumps, regulators, and storage or treatment units for water conveyance towards outfalls (Fig. 2.2). Surface runoff, groundwater interflow, and sanitary dry weather flow act as inflow. The outfall undergoes the available treatment facilities.

According to U.S. Environmental Protection Agency (US EPA) (USEPA 2007) for a storm event:

- Natural ground cover: could contribute 40% as evapotranspiration, 10% in the runoff, while 25% shallow infiltration and rest 25% deep infiltration.

Fig. 2.2 Typical urban drainage system

- 10–20% impervious surface: might lose 38% as evapotranspiration and 20% runoff while 21% shallow infiltration and rest 21% for deep infiltration.
- 35–50% impervious surface: would lose 35% as evapotranspiration and 30% runoff with 20% shallow infiltration and 15% deep infiltration.
- 75–100% impervious surface: contribution as evapotranspiration could be 30%, then 55% runoff with 10% shallow infiltration and only 5% deep infiltration.

Thus, the typical urban drainage system comprises a combined sewer system and various drains to dispose of stormwater and urban wastewater. There might be individual or combined arrangements (Fig. 2.2). *Stormwater* is originated from precipitation then passed over a land surface before being disposed of toward a natural water body. This stormwater could be stagnant on the impervious surface and slowly disposes into nearby drains, channels, streams, rivers, and eventually the ocean (Fig. 2.2).

Contrary, the water smart city is designed to overcome the associated water-related challenges:

- Alteration of the stream flows;
- Alteration of stream channels alignments and the related ecosystem;
- Increased siltation and sedimentation; and
- Degraded water quality.

Therefore, the impacts could be quantified by adequately inspecting the hydrological cycle, water regime, and water quality of adjacent urbanized territory. Excess rainfall/ stormwater harvesting might offer the Best Management Practices (BMPs) to minimize stormwater runoff, and the associated problems are described in Chap. 4.

2.4 Precipitation

Precipitation is the prime link in the water cycle to convey atmospheric water to the earth. The required airlifting mechanisms to form precipitation are classified into three categories: i.e., *orographic lifting*: is caused or enhanced by one or more of the effects of mountains on the earth's atmosphere, *frontal surface lifting*: warmer air is lifted above cooler air towards equilibrium with a more cool surface; finally, for the *convective lifting*: the upward movement of warm air from a warm surface towards progressively clam down. The different forms of precipitation are snow, hail, sleet, glaze, fog, smog, dew, mist, drizzle, and rain. Snow, hail, sleet, and glaze are experiences at low temperatures. *Fog* is a low-level cloud that touches the ground, and *Mist* is the floating/falling water particle in the atmosphere at or near the earth's surface that turns to rain. *Drizzle* is the rain in tiny light drops. Most of the precipitation falls as rain. Based on the size and velocity, rains are categorized into Light rain, Moderate rain, Heavy rain, Excessive rain, and Cloudburst. Cloudburst is a rainfall of exceptionally very high intensity. High-intensity precipitation occurs over a substantial time; covering a large area is generally termed as a storm. Table 2.1 presents the physical properties of different types of precipitation.

Table 2.1 Precipitation size and speed (Lull 1959)

	Intensity (mm/h)[a]	Median diameter (mm)	Velocity of rainfall (m/s)	Drops per sec/sq.m.
Fog	0.13	0.01	0.003	67,425,000
Mist	0.05	0.1	0.21	27,000
Drizzle	0.25	0.96	4.1	151
Light rain	1.0	1.24	4.8	280
Moderate rain	3.8	1.60	5.7	495
Heavy rain	15.2	2.05	6.7	495
Excessive rain	40.6	2.40	7.3	818
Cloudburst	102	2.85	7.9	1,220

[a]*Rainfall intensity* is the accumulated rainwater depth on a surface divided by the rainfall duration, expressed in mm/h.

The World Meteorological Organization (WMO) defines *Probable Maximum Precipitation* (PMP) as "the greatest depth of precipitation for a given duration meteorologically possible for a given size storm area at a particular location at a particular time of year, with no allowance made for long-time climatic trends" (WMO 1986).

2.4.1 Measurement of Rainfall

Rainfall is generally measured using rain gauges by meteorologists and hydrologists. Two types of rain gauges are available, i.e., Non-recording or storage rain gauges and Recording rain gauges. With *non-recording rain gauges,* total rainfall is measured over a day or longer, and shorter-time rainfall events are by *recording rain gauges*. Consequently, recording rain gauges measure both the rainfall amount and its duration.

- Non-recording rain gauges

Non-recording rain gauges collect and store rainfall over a period, usually daily (Fig. 2.3). For the daily gauges, readings are typically taken once a day at 9.00 a.m. by transferring the contents of the gauge into a particular measuring cylinder shown in Fig. 2.3a. The maximum capacity of the measuring cylinders is 10 mm rainfall. Once emptied, the gauge returned in position and was ready to collect for the next 24 h. Apart from a daily rain gauge, the U.S. National Weather Service (NWS) has used a standard rain gauge in cumulative rainfall measurements for more than 100 years (Fig. 2.3b). Information provided by storage gauges is often too coarse for some applications, e.g., flood forecasting and urban drainage design. For these applications, rainfall data over shorter time intervals or real-time data are preferable, and discussions on advanced technologies are included in Chap. 8.

- Recording rain gauge

These gauges record both the amount and duration of rainfall; hence, they are helpful in applications for which the rainfall intensity over very short durations is essential. There are three types of recorders in use:

Weighing bucket-type rain gauge is a self-recording rain gauge comprised of a receiver bucket on spring support or lever balance. Bucket moves due to accumulative self-weight and conveys a pen mark on a clock-driven chart (Fig. 2.4a). This rain gauge plots the cumulative rainfall depths against time, and the plotted curve is known as the *mass curve*.

Tilting-siphon (float-type) rain gauge follows a working principle similar to weighing bucket rain gauge. An inlet funnel (A) collects the rainfall forwarded to a collecting chamber (B). The upward movement of the float (C), which provides a measure of the rain, is recorded by a pen (D) on a standard graph paper mounted on a rotating drum (Fig. 2.4b). Thus, the pen records the cumulative amount of

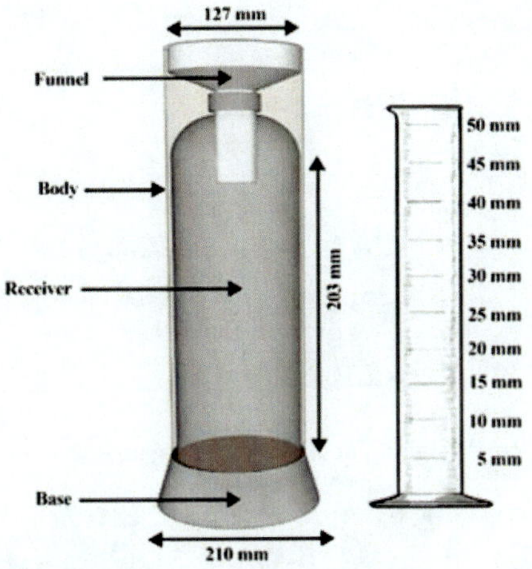

(a) Symons non-recording type rain gauge

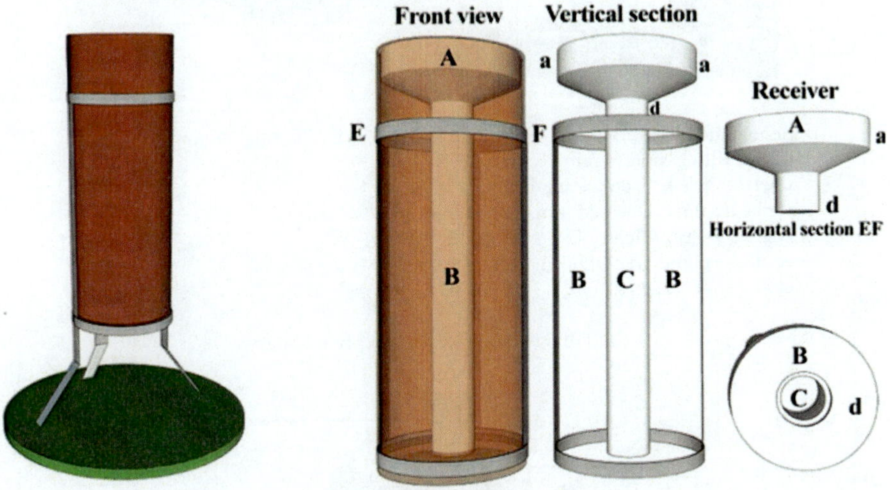

(b) Standard rain gauge: U.S. Weather Bureau pattern 200 mm copper rain gauge

Fig. 2.3 Non-recording rain gauges

rainfall against time. The chamber fills up, the pen lifts off the chart top leading to the activation of a siphon and empty the chamber. The main problem with the tilting siphon is that it may miss receiving the rain during its emptying. Thus, although the amount of uncaught rain resulting from this is usually small, it is essential to have a storage gauge nearby whose measurement can be used to adjust the reading of the tilting-siphon if necessary. Additionally, post-processing of the charted rainfall is required to develop rainfall intensity-duration relations. This post-processing usually

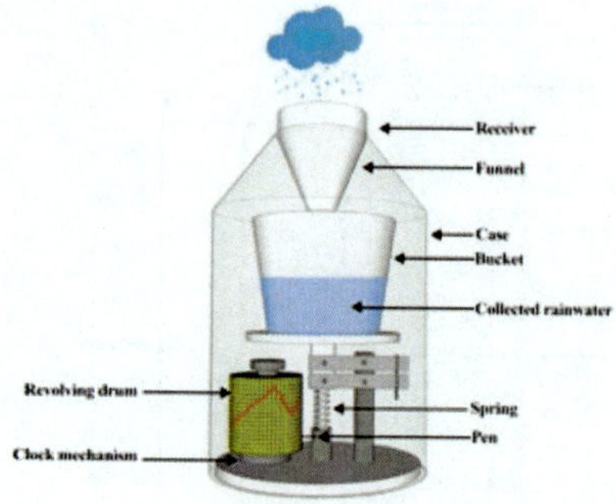

(a) Weighing bucket-type rain gauge

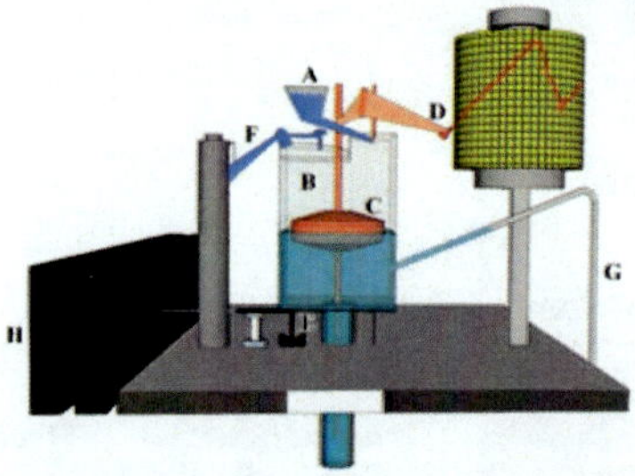

(i) Schematic of the working mechanism of the tilting-siphon rain recorder. A = inlet funnel, B = collecting chamber, C = plastic float, D = pen arm, E = knife-edge, F = trip release, G = siphon, H = counterweight. (reproduced from Shaw et al. (Shaw *et al.*, 2011)

(ii) Tilting-siphon rain recorder

(b) Tilting-siphon (float-type) rain gauge

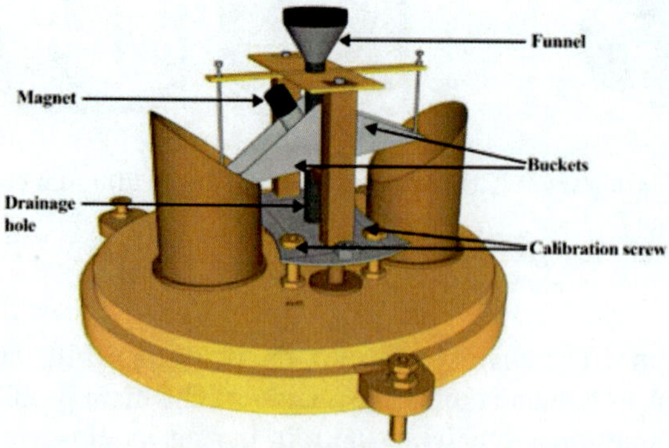

(c) Tipping bucket rain gauge

Fig. 2.4 Recording rain gauge

involves manually extracting rainfall depths of various durations from the charts, resulting in significant errors.

Tipping bucket rain gauge breaks an entire storm into a series of events, unlike the above-mentioned two rain gauges, hence the name event recorder. First, one of the buckets is opened immediately below the funnel, which collects the rain. Once the bucket is complete, the water weight causes it to tilt away from the funnel and empty; this is tipping. The second bucket is now positioned immediately below the funnel to receive the following 1 mm of rainfall, and the process continues until the storm stops (Fig. 2.4c). Meanwhile, a magnet mechanism attached to the supporting rod for the buckets triggers a switch at the bottom of the equipment each time tipping occurs. By doing so, the time between the tippings, i.e., the duration of each 1 mm of rainfall, is recorded. The more intense the storm, the shorter is the duration of a 1 mm rainfall. Collected information by the tipping-bucket data represented in *hyetographs* (or a plot of precipitation against time).

- Other Recent Rainfall Measurement Techniques

RAdio **D**etection And **R**anging (RADAR) technique belongs to a large group called remote sensing. RADAR and other methods are used to measure rainfall, covering a single point to the entire catchment. RADAR, therefore, gives information about the spatial distribution of rainfall during an event. RADAR technique is costly and also requires a comprehensive network of conventional rain gauges to calibrate. However, an extensive database of radar-collected precipitation is available around the world.

- Rain gauge network

Two types of errors are recorded in rainfall measurement, i.e., random errors and systemic errors. *Random errors* are caused by rain gauge density, storm characteristics, and the rain gauge capability to represent the area as per design. *Systematic errors* have resulted from measurement, improper placement, inadequate exposure, change in the observer and gauge.

Generally, the rainfall sampling errors increase the proportion to the mean areal rainfall and are inversely proportional to the network density, prolonged rainfall duration, and areal extent. Hence, a specific network for storm rainfall produces more significant average errors than for periodic rain, i.e., monthly, seasonal or annual. The statistical assessment is carried out on an existing rain gauge network within a watershed. In the estimation of mean areal rainfall, the optimum number of rain gauges (N) is as:

$$N = \frac{C_v}{\epsilon} \tag{2.1}$$

where

C_v = Coefficient of variation of the recorded rainfall by the gauges

ϵ = Assigned percentage of error in the predicted mean areal rainfall (usually 10%).

According to the WMO, the recommended rainfall network density should follow:

- Flat regions, Mediterranean and Tropical zone: One station per 600–900 km^2.
- Mountainous regions, Mediterranean and Tropical zone: One station per 100–250 km^2.
- Small mountainous land with uneven precipitation: One station per 25 km^2.
- Arid and polar zones: One station per 1,500–10,000 km^2.

2.4.2 Rainfall Data Processing and Presentation

The raw rainfall numbers (or data) collected from a gauge need to be analyzed to derive useful information. Analyses of point rainfall data might include estimating missing rainfall at several gauges, checking the consistency of rainfall data at gauges, estimating catchment rainfall from rainfall at individual gauges, and then developing relationships between rainfall intensity and its duration.

- Missing data is undoubtedly familiar in countries where data are still manually collected. Unsuitable data exist but are doubtful because they are not compatible with the rest, either too low or too high. This class of data is called *outliers*. Whatever the circumstance, it is often beneficial to infer the missing or unsuitable data for a particular gauge, using data at other gauges with complete data. A straightforward method of assuming missing rainfall data is the normal-ratio method (Linsley et al. 1992).
- The *double mass* method performs for the consistency check of hydro-meteorological data records. This technique is based on the fact that, while the accumulated mass of rainfall for several gauges is not very sensitive to changes at the individual sites, the accumulated mass of rain at any component gauges is susceptible to changes at that site. Thus, the plot at any single location should be a straight line: any departure from linearity indicates change or inconsistency that should be removed. Therefore, this change is presented by a deviation in the slope of the straight line.
- Rainfall data recorded at the individual gauges within a catchment depends on the event's reason and nature. Thus, to understand the catchment responses, rainfall in any duration has converted to an equivalent for the entire catchment. Typically, four methods are applied for catchment rainfall estimation: Arithmetic mean method, Thiessen polygon, Isohyetal method, and Hypsometric method.

The *arithmetic mean method* applies to determine rainfall depth over the catchment as the average of the individual gauge measurements, i.e.

$$\overline{P} = \frac{\sum_{i=1}^{n} P_i}{n} \tag{2.2}$$

where $\overline{P}$ is the average rainfall depth over the catchment (mm); P_i is the rainfall depth (mm) at i th gauge, and n is the total gauge numbers.

Contributing area-weighting methods assess each point measurement by its contributing area while estimates catchment rainfall. There are two techniques by which these contributing areas have been determined are Thiessen polygon method and the Isohyetal method. *Thiessen polygon method* assigns aerial impacts to point rainfall values. Here, bisectors are perpendicularly constructed by joining the representative lines from each station. Then, the bisectors form a series of polygons, and each polygon containing one station. Rainfall values recorded at a station is assigned to the entire area covered by the enclosing polygon:

$$\overline{P} = \frac{\sum_{i=1}^{n} P_i a_i}{A} = \sum_{i=1}^{n} P_i \left(\frac{a_1}{A}\right) \tag{2.3}$$

where a_i is the contributing area (or Thiessen polygon area) for gauge i th, km^2 and A is the total catchment area, km^2.

Conversely, the *Isohyetal method* draws contours of equal rainfall depths (or isohyets) using values recorded at the individual gauges.

$$\overline{P} = \frac{\sum_{i=1}^{N} p_i a_i}{A} \tag{2.4}$$

where pi is the mean rainfall (mm) between two isohyets, a_i is the area enclosed between two isohyets km^2 and N is the number of these areas.

The *Thiessen approach* gives poor results for convective and orographic storms. The isohyetal method performs better within a catchment in an active orographic influence. The main problem with the isohyetal approach is that the map should always be drawn for each rainfall event. Even the area-weighting methods would give erroneous results when a catchment is topographically diverse.

The *Hypsometric method* developed by the WMO (1986) enables topography to be considered when estimating average catchment rainfall. A basic assumption of the technique is that a linear relationship exists between rainfall magnitude and station altitude.

$$\overline{P} = \frac{V}{A} \tag{2.5}$$

where P is the precipitation (mm) over the catchment

V is the volume of water under the curve of precipitation (m^3).

2.4.3 *Rainfall Depth-Area-Duration*

Once sufficient rainfall records for the region are collected, the primary or raw data could be analyzed and processed to produce useful information in curves or statistical values for planning and development of water resources projects. *Depth-Area-Duration* (DAD) analysis of a storm applies to determining the maximum amounts of rainfall within various durations over different areas. The DAD curves can be obtained through the following steps:

- Examine the rainfall records of the region in which the catchment area is under consideration. Also, consider forms of meteorologically similar regions. From it, prepare a list of the most severe storms with their dates of occurrence and duration.
- For the listed severe storms, prepare isohyetal maps and determine the rainfall values over the area of each isohyet (rainfall contour).
- Plot a graph curve connecting area and rainfall values for different durations, say 1 Day rainfall, 2 Day rainfall, 3 Day rainfall (Refer to Fig. 2.5).

Technical Paper 29 (TP-29) is the first technical paper by the U.S. Weather Bureau that attempted to assess DAD relationships (USDA 1957). With their limited computing resources, however, their method of choice was simply to take the arithmetic mean of station recordings:

The estimation of areal rainfall with a sufficient volume of data to derive general regional duration and frequency relationships could become so laborious as to defeat its purpose. With no precedent for this work, it was necessary to test methods for processing the data. However, DAD relationships only extend to 1000 km^2. Technical Paper 40 (TP-40), titled "Rainfall Frequency Atlas of the US for durations from 30 min to 24 h and return periods from 1 to 100 years" (USDA 1961), was the first rainfall frequency atlas in the US. In TP-40, statistical stationarity was assumed

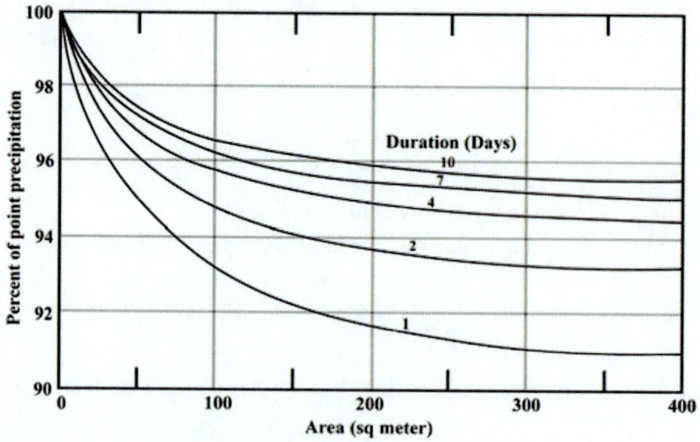

Fig. 2.5 The depth-area-duration relationship (USDA 1961)

throughout the record, 1938–1957, to determine frequency rainfall depths. Since then, TP-40 was an excellent resource for hydrologic studies and was used nationwide until NOAA Atlas 14 superseded it in some regions.

Most severe storms in the listed storms may not have occurred right over the catchment under consideration, but there is a possibility of such occurrence. So, from DAD curves, 1 Day, 2 Day, 3 Day rainfall depths for a catchment area of the proposed project needed to record. These provide the rainfall depths when the storms experience over the catchment.

2.4.4 Rainfall Intensity–Duration–Frequency (IDF)

The rainfall *intensity–duration–frequency* (IDF) relationship offers fundamental importance in the water resources design. Here, rainfall duration has divided into three divisions, i.e., short: 1 min to 1 h, intermediate: 1–24 h, and long: more than 24 h. For rainfall P (mm) falling over duration D (hours), an average rainfall intensity *I* is:

$$I\left(\frac{mm}{h}\right) = \frac{P}{D} \tag{2.6}$$

The frequency (F) of a storm of a given intensity is the probability that a given rainfall intensity will be equaled or exceeded within a given number of years. Then, the outcomes of the frequency analysis plot on the log–log paper. Typically, F is expressed as the return (or recurrence) period (Fig. 2.6). The *return period* (in years) is the inverse of the probability of exceedance. It gives the average interval between a given random event and another event of a similar or higher magnitude. For example, return periods for different structures are (i) 2 years for urban drainage, (ii) 5–10 years for field structures, (iii) 20 years for gully control and small farm dams, and (iv) 50 years for large farm dams. *Mild* (intensity) storms that occur relatively frequently have a high probability of being equaled or exceeded, i.e., short return periods. High-intensity, flood-generating batteries have a minimal probability of being equaled or exceeded, i.e., long return periods. The probability of exceedance (and hence the return period, T) for a given storm intensity requires statistical analysis of annual maximum storm intensities of various durations extracted from records of rainfall recorders over several years. Thus, two options are available i.e.

- *An empirical approach* is applied to plot the probabilities exceedance based on the observations.
- *The theoretical approach* includes the observations using Generalized Extreme Value (GEV) distribution, Gumbel (EVI) distribution, Gamma distribution, Log Pearson III distribution, Lognormal distribution, Exponential distribution, and Pareto distribution. These are used to estimate the rainfall events for a given exceedance probability.

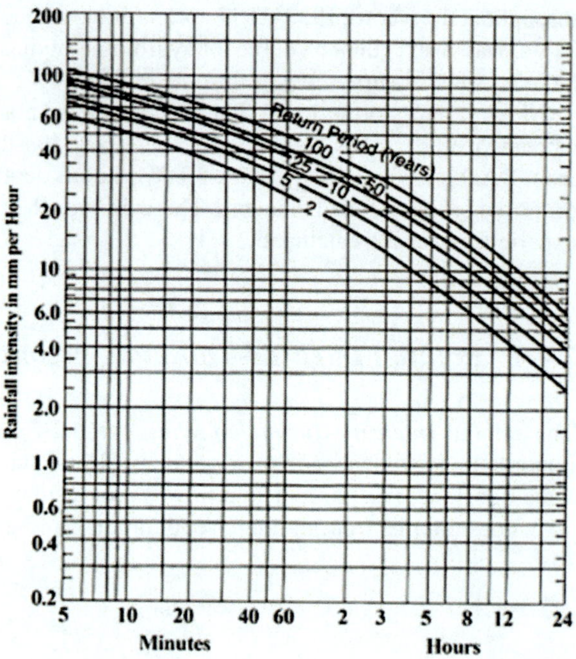

Fig. 2.6 Typical IDF schematic

The empirical approach uses the standard formulas are Sherman's formula (1931), Chow (2009), and several procedures are available; many are site-specific. According to Sherman's (1931) formula:

$$\text{Rainfall intensity (mm/h)}, I = \frac{\text{K}a}{(D + c)^e} \tag{2.7}$$

where

D = Rainfall duration (min)

K, a, c and e = the constant parameters related to the metrological conditions; these are the function of the return period.

Also, Chow (2009) described as:

$$\text{Rainfall intensity (mm/h)}, I = \frac{K F^m}{(D + b)^n} \tag{2.8}$$

Here, K, b and m, n = coefficient, constant, and exponents, respectively, depending on conditions that affect rainfall intensity.

Developing and using IDF curves are influenced by geographical locations and rainfall patterns, and distributions. Thus, the National Oceanic and Atmospheric Administration (NOAA) Atlas #14 provides IDF curves and constructs design storms for the USA. In the UK, this task has followed from the Flood Estimation Handbook (Reed 1999); similar studies have also been found worldwide. The general empirical

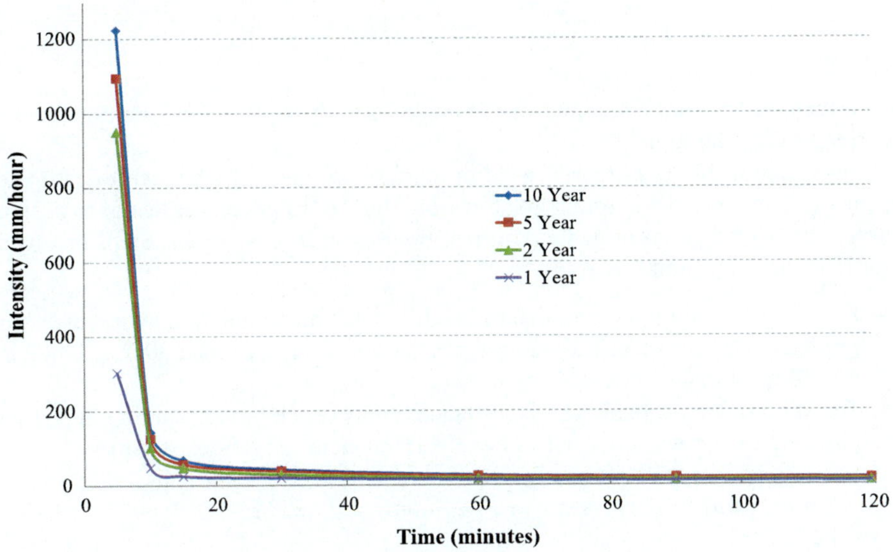

Fig. 2.7 IDF curves for different frequencies

relationships are presented in Fig. 2.7. Thus, for a fixed F, I decreases with D as follows:

$$I = \frac{a}{D + b} \; or \; I = \frac{c}{D^n} \tag{2.9}$$

where a, b, c, and n are the empirical constants, those can be determined by analyzing data at specific locations.

Gumbel (Type I) distribution is the typical approach for the theoretical *Extreme Value (E.V.)* Distribution. The Gumbel Type I distribution expresses:

$$G(P; \mu, \beta) = \frac{1}{\beta} e^{\frac{x-\mu}{\beta}} e^{-e^{\frac{x-\mu}{\beta}}} \tag{2.10}$$

where μ is the location parameter, and β is the scale parameter.

Mathematically, the precipitation (P) is directly related to the average intensity I and the duration D, the associated random variable X_T over a given return period (T) might be obtained from:

$$X_T = \overline{X} + K_T \sigma \tag{2.11}$$

where $\overline{X}$ is the arithmetic average of the observed record, and σ is the standard deviation of the observations. The frequency factor (K_T) associated with T is given by:

$$K_T = -\frac{\sqrt{6}}{\pi}\left[0.5772 + \ln\left(\ln\left(\frac{T}{T-1}\right)\right)\right] \tag{2.12}$$

Equations (2.10), (2.11), and (2.12) apply to each set of annual maxima to the corresponding duration.

IDF relationship is primarily used in the rational method approach to estimate peak discharge resulting from a given storm. This information is essential to design urban drainage systems or in the design of flood protection schemes. Steps involved for IDF curve preparations are:

- From yearly rainfall records, determine the annual maximum rainfall intensity (or rainfall depth) for specific durations, i.e., 5 min, 10 min, 15 min, 30 min, 1 h, 2 h, 6 h, 12 h, and 24 h.
- To estimate the probabilities exceedance, apply an empirical plotting approach based on the observations. Calculate the frequencies of various storms:

$$Total\ number\ of\ years = Frequency \times Ranking\ of\ storm \tag{2.12a}$$

- Fit a theoretical EV distribution (e.g., Gumbel Type I) over the observed records to estimate the rainfall events associated with the given exceedance probabilities.

Example Problem 2.1 A historical database at a station for 40 years was analyzed, and ten extreme events were stipulated in their decreasing order, as shown in Table 2.2. Plot Intensity–Duration–Frequency (IDF) curves.

Solutions

Four different frequencies are selected to plot the IDF curves, i.e., 10 years, 5 years, 2 years, and 1 year.

Table 2.2 Computation for Example problem 2.1

Record no	5 min	10 min	15 min	30 min	60 min	90 min	120 min	Frequency[b] (years)
1	102	24	16.8	20.88	25.8	29.52	35.64	10.0
2	91.2	20.8	14.16	18.6	23.04	28.56	31.56	5.0
3	87.6	18.6	13.32	16.32	20.4	25.68	28.08	3.33
4	86.4	17.6	12.36	14.64	17.4	21.72	25.44	2.50
5	79.2	16.8	11.64	14.16	16.8	19.8	21.96	2.00
6	74.4	16	11.04	13.2	15.96	18	19.68	1.67
7	61.2	15.6	10.8	12.6	15	16.8	18.6	1.43
8	54	13.6	9.84	12.12	14.4	16.32	18.12	1.25
9	43.2	10.4	8.04	11.4	13.68	16.08	17.52	1.11
10	33.6	10.2	7.44	9.96	13.32	15.24	16.92	1.0

[b] Calculated

Table 2.3 Computation of the average precipitation intensity (mm/hour)

Duration (min)	Average intensity mm/h[c]			
	10 years	5 years	2 years	1 year
5	1224	1094.4	950.40	302.4
10	144	124.8	100.80	46.8
15	67.2	56.64	46.56	24
30	41.76	37.2	28.32	18.96
60	25.8	23.04	16.80	13.08
90	19.68	19.04	13.20	9.84
120	17.82	15.78	10.98	8.04

[c] Calculated based on Table 2.2

Based on Table 2.3, IDF curves can be plotted for these four different frequencies, as shown in Fig. 2.7.

2.5 Infiltration

Precipitation on lands subsequently enters the soil due to gravity and capillary action, and this process is called *infiltration*. The infiltration process has illustrated in Fig. 2.8. Under ponded conditions, there are five zones involved in the idealized homogeneous soil profile for the infiltration process:

- **Saturated Zone**: contains water-filled pore spaces. Based on the elapsed time since the initial water application, this zone extends only a few millimeters.

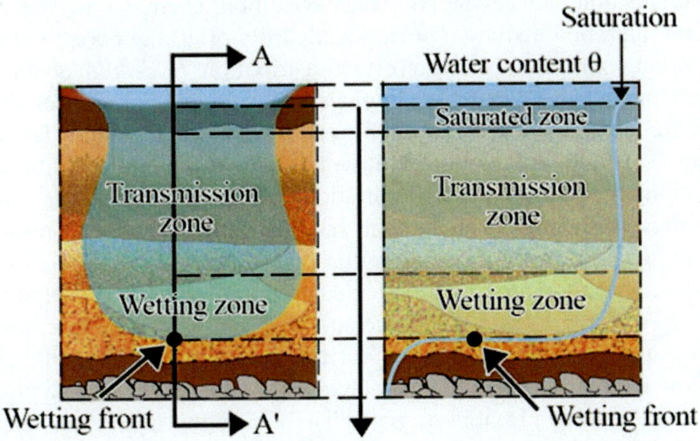

Fig. 2.8 Schematic of infiltration processes (adapted from Hillel 2013)

- **Transition Zone**: contains a quick declination in moisture/ water content with depth, and this zone extends few centimeters.
- **Transmission Zone**: contains unsaturated zones with uniform moisture content, making a minor change in water content due to gravitational forces.
- **Wetting Zone**: consists of lessened water content than the transmission zone and abrupt reductions with depth similar to the initial moisture content of the soil.
- **Wetting Front** consists of a steep hydraulic gradient of metric potentials and a sharp boundary between wet and dry soil. Beyond this zone, penetration of water is usually absent.

Infiltration replenishes the soil profile, and the excess water moves downward by gravitational force, i.e., *seepage* or *percolation,* and recharges towards the groundwater table. If the water supply to the soil surface exceeds the absorption capacity of the soil, then the excess water becomes *runoff*. The maximum water-absorbing rate of the soil is the *infiltration capacity* or *infiltrability.* Infiltrability estimates water availability for percolation based on the information on runoff, drainage, or evapotranspiration. Generally, considering ponding conditions over the soil surface, the infiltration rate exceeds the soil infiltrability. Based on the speed of precipitated water applications on soil, the infiltration process has been classified into two (Hillel 2013), i.e., supply controlled and profile controlled. Due to slow or low precipitation infiltration rate, the *supply controlled* infiltration rate becomes lower than the soil infiltrability. *Profile controlled* infiltration process handled higher infiltration rates exceeding soil infiltrability, i.e., the *actual infiltration rate*.

Infiltration rate is affected by soil's chemical and physical properties, LULC, topography, precipitated water properties, existing groundwater table, and the type of equipment or method used to ensure infiltration. Thus, the leading factors are flow influences, soil surface conditions, hydrophobicity, subsurface conditions, and root system. On the other hand, surface and subsurfaces can affect infiltration. These are mechanical processes, plowing, frost-freeze–thaw cycles, litter layer, organic content, compaction, antecedent soil water condition, chemical activity, biological activity, and microbial activity. *Porosity* is the ratio of all the pores' volume to the total soil volume, influencing soil moisture distribution. *Hydraulic conductivity (K)* is a property of soils that ensures water movements through pore spaces or fractures. A list of K for the various soils is available in Appendix B, Table B.1. Due to a lower suction gradient, initially saturated soil would have lesser initial infiltrability and, after that, rapidly reaches a constant infiltration rate. *Saturated hydraulic conductivity* (Ksat) ensures water movement through saturated media, and *Ksat* proportionally increases infiltrability within the soil profile (Table B.1).

Naturally formed soil profiles are usually non-homogeneous with depth; comprise separate layers with particular hydraulic and physical features. These layers within the soil profile delay water movement during infiltration; for example, clay layers obstruct flow with lower K_{sat}. But, clay layers near the surface cause a higher initial infiltration rate and then rapidly drop off. The infiltration rate is influenced by the disturbance of the softened sediment surface. Table 2.4 describes the variation of infiltration rates for different types of soils. The effect of rainfall on bare soil might have

Table 2.4 Infiltration rates for different type soils (Johnson 1963)

Soil type	Porosity (%)		Infiltration rate (mm/h)
	Total	Non-capillary (specific yield)	
Gravelly silt loam	54.9	28.1	126.0
Clay loam	61.1	36.3	101.1
Silt loam	57.0	32.0	53.1
Sandy loam	49.6	26.3	49.0
Clay (eroded)	54.3	28.7	45.2
Sandy clay loam	48.8	27.7	36.1
Silty clay loam	50.8	24.3	18.3
Stony silt loam	59.7	32.6	14.0
Fine sandy loam	41.5	24.2	14.0
Very fine sandy loam	49.6	23.4	12.9
Loam	45.7	17.2	12.7
Sandy clay	42.9	16.9	1.3
Heavy clay	57.8	27.0	0.5
Light clay	47.0	19.8	0.0
Clayey silt loam	49.4	17.6	0.0

a considerable influence over vegetation on the infiltration rate (Wisler and Brater 1959). Plant roots zone also intensifies infiltration by upward hydraulic conductivity of the soil surface. As stated in Sect. 2.2, the infiltration remains to decrease with increased urbanization, resulting in groundwater lowering. Apart from the different types of LULC, topography has an indirect impact on the infiltration rate. For instance, steep slopes would increase the runoff and thus influences the required time for infiltration.

In contrast, gentle slopes with decreased runoff having lower infiltration. Therefore, infiltration affects water's chemical and physical properties, applied water head, time length of water application, biological activity, water temperature, and the sediment's entrapped air percentage. The hydrologic soil groups as per the Natural Resources Conservation Service (NRCS) soil scientists are:

Group A: Soils with low runoff and high infiltration rate, when thoroughly saturated, deep, well-drained sands or gravels.

Group B: Soils with a moderate infiltration rate when fully saturated, viz. moderate deep to deep, moderately well to well-drained soils, with moderate fine to moderate coarse soil texture.

Group C: Soils with a slow infiltration rate when fully saturated comprise a soil layer that impedes downward water movement or is moderately fine to a fine soil texture.

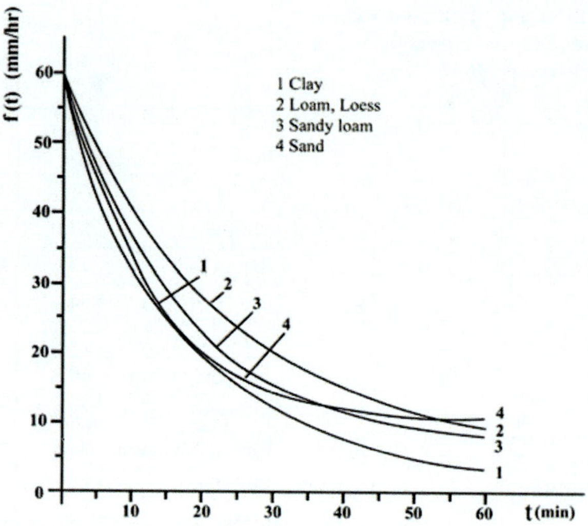

Fig. 2.9 Variations of infiltration capacity following Critchley and Siegert (2013)

Group D: Soils with a high runoff and slower infiltration rate when fully saturated, viz. high swelling potential clay, soils with a permanent high water table, soils with a claypan or clay layer at or near the surface, and shallow soils over nearly impervious material.

The *infiltration capacity* for a given soil is the maximum infiltration rate at a given time. Starting from a storm event, the high infiltration capacity of the soil generally observes and exponentially decays as time elapses (Fig. 2.9). Thus, the actual rate of infiltration (f) expresses as:

$$\begin{aligned} f &= f_p, \text{ when } I \geq f_p \\ f &= I, \text{ when } I < f_p \end{aligned} \tag{2.13}$$

where

I = rainfall intensity

f_c = infiltration capacity (mm/h).

Soil properties and LULC greatly influence infiltration capacity; thus, the soil of higher infiltration capacity could facilitate groundwater storage. On the other hand, forests with organic-rich soil would have more infiltration capacity than paved urban areas.

2.5.1 Measurement of Infiltration

Infiltration data can be obtained from three methods, i.e., (i) Field testing, (ii) Soil grain size analysis, and (iii) Hydrologic equation for an area.

Field testing

This testing applies with infiltrometer; based on their working principle, infiltrometers are in three different forms, i.e., disk permeameters, ring infiltrometer, and drip infiltrometer. Observations from infiltration pits and ponds also practice worldwide, and their design is based on the drainage basin size.

Disc permeameter or tension infiltrometer is widely used to obtain the soil's infiltration rate and hydraulic conductivity. Thus, the soil characterization is based on the saturated and unsaturated soil hydraulic properties. Two types of ring infiltrometers are available, i.e., single ring infiltrometer and double-ring infiltrometer.

In *Single Ring Infiltrometer*, single steel ring known boundary and pressure conditions confine soil below the ring. A constant water head is supplied manually or from marionette bottles to determine the soil's free 3D infiltration capacity (Fig. 2.10a). Then, soil permeability is measured through steady-state calculations and Ksat computation. The infiltration rate is the amount of water that infiltrates into the soils per surface area, per unit time. The typical diameter of the single ring is 305 mm, or 457 mm, or 610 mm. The *Double Ring Infiltrometer* consists of an inner and outer ring inserted into the ground to measure saturated hydraulic conductivity with a similar work principle. K_{ast} can be estimated for the soil when the water flow rate in the inner ring is steady-state. This ring allows much easier calculation for lateral flow. Hence, a double ring infiltrometer is a widely used instrument. Figure 2.10b schematically states the vertical flow of a double ring infiltrometer generated from the inner ring; this is allowed by the outer ring for all the lateral flow.

Typical diameters of the double ring infiltrometer are any of the three combinations, i.e., 305 and 457 mm, 305 and 610 mm, and 457 and 610 mm. *Rain simulators* or *drip infiltrometers* are widely used in infiltration and runoff studies by applying water to the soil surface with specific energy, thus replicating field conditions during rain showers. This infiltrometer brings about a splash effect that may reduce macroporosity in the soil surface and reduce hydraulic properties during the measurements. Larger scale measurements using rainfall simulators often measure variability within a plot due to LULC and topography and relate to the entire field's lumped infiltration rate. *Observation from infiltration pits and ponds* has been carried through deducting the loss due to evaporation. To measure the infiltrating waters at that point, a catch basin, i.e., a lysimeter placed under a laboratory sample or at some depth below the land surface. A *lysimeter* accurately measures infiltration once a large soil tank is considered for vegetation. This tank allows rainfall as input and water lost through the soil by evapotranspiration. Most studies on the spatial variability of infiltration consider point measurements through ring infiltrometers or disk permeameters. These cost-effective and time-saving field test measurements offer a direct infiltration rate. Field tests might be used to determine the feasibility of bioretention, permeable pavement, or rain gardens to fulfill minimum regional or country requirements. Details on bioretention, permeable pavement or rain gardens are available in Chap. 4.

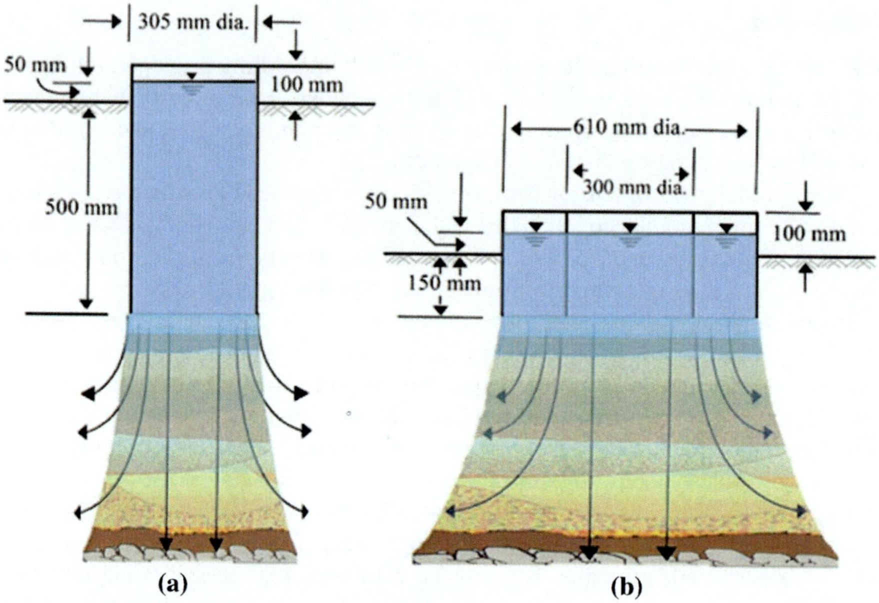

Fig. 2.10 Infiltrometer

Soil Grain Size Analysis

Infiltration rates in a catchment are determined through soil gradation and textural analysis. Also, differences between soil moisture and water table rise might provide the infiltration rates. Soil grain size analysis is practiced for the different stormwater management approaches (details are in Chap. 4). Worldwide different standards are available for soil sampling and grain size analysis; the typical recommendations (Kale et al. 2019) are:

- For the infiltration basins and trenches, the grain size analysis for each defined layer below the infiltration facility should be carried out to 2.5 times the maximum ponding depth, but not less than 2 m.
- For bioretention areas and permeable pavement, the grain size analysis is to be performed for the layer beneath the top of the final bioretention area/the last subgrade to a depth of 3 times the maximum ponding depth not less than 1 m.

The *hydrologic equation for an area* or *Hydrograph Analysis* and the corresponding rainfall records provide a small watershed infiltration capacity. Based on adequate rainfall and runoff records conforming to isolated storms in a small watershed, water abstraction with relatively homogeneous soils could be estimated using the water-budget equation.

2.5.2 *Infiltration Models*

The objective in modeling infiltration is to quantify the infiltration rate from a given rainfall in a given time interval. This model has dealt with differential equations governing the water flow in unsaturated porous media to acquire constant infiltration rates. Two types of equations are available to calculate infiltration, i.e., *empirical equations* and *physically-based equations*.

Empirical equations

The predictable and well-shaped infiltration capacity curves or relationships between infiltration rate and time acquire through physical significance to the empirical model parameters. The Lewis-Kostiakov and Horton (Horton 1940) equations widely use among the available empirical equations. The empirical expression for the Lewis-Kostiakov equation independently proposed by Kostiakov (1932) and Lewis (1937), based on the curve fittings on field data. This empirical equation provides infiltration to time as a power function:

$$\text{Infiltration capacity, } f_p = K_k t^{-\alpha} \tag{2.14}$$

where

$t =$ the elapsed time

$K_k, \alpha =$ empirical parameters.

Lewis-Kostiakov equation is widely used due to simple computations of the two constants from measured infiltration data and reasonably fit infiltration data for various soils over a given time. The limitations involve:

- Infiltration capacity prediction is infinite at t equals zero, and for long infiltration times, it erroneously predicts zero rates instead of a steady value.
- This infiltration capacity fails to be adjusted for different field conditions and soil moisture content that adversely affect infiltration.

The *Horton Equation* provided by Horton (1939, 1940) was empirical in an exponential form. Here, infiltration capacity (f_p) decreased with time until reached the minimum constant rate (f_c). Thus, the infiltration capacity (mm/hour) explicitly as a function of time:

$$f_p = f_c + (f_o - f_c)e^{-kt} \tag{2.15}$$

where

$f_o =$ Initial infiltration rate (mm/h)

$f_p =$ Infiltration capacity (or maximum infiltration rate, mm/h)

$f_c =$ Final constant infiltration rate (mm/h) as $t \to \infty$

$t =$ Time since the start of rainfall (min)

$k =$ The empirical coefficient for a particular LULC (min^{-1}).

Table 2.5 Estimated parameters for Horton's infiltration model (Wilson 1983)

LULC		f_o (mm/h)	f_c(mm/h)	k (min^{-1})
Standard agricultural	Bare	280	6–220	1.6
	Turfed	900	20–290	0.8
Peat		325	2–29	1.8
Fine sandy clay	Bare	210	2–25	2.0
	Turfed	670	10–30	1.4

If there is vegetation, k is small, whereas a smooth bare soil that is less porous will have a higher k. Both f_0 and f_c also depend on soil properties and land use. Typical values for these correlations are tabulated in Table 2.5.

To find infiltration using the Horton model is simply by integrating the Horton equation over the desired period. The assumption is, infiltration will always proceed at the infiltration capacity rate and that this rate will always be lower than the rainfall intensity. If this is not the case, the infiltration will be over-estimated and under-estimating the rapid response of runoff. Consider the hyetograph below on which the Horton curve is superimposed (Fig. 2.11). However, rainfall intensity could be less than the infiltration capacity predicted by the Horton model. On those occasions, assuming that infiltration proceeds at the capacity are wrong, infiltration can only occur at the rainfall rate, as written in Eq. 2.16. The direct integration method is not advised; instead, a sequential approach in which each rainfall block interval of the hyetograph is considered in turn:

$$f(t) = minimum\left[f_p(t), I(t)\right] \tag{2.16}$$

In Fig. 2.11,

$\Delta P_e =$ Depth of excess rainfall in interval Δt (mm)

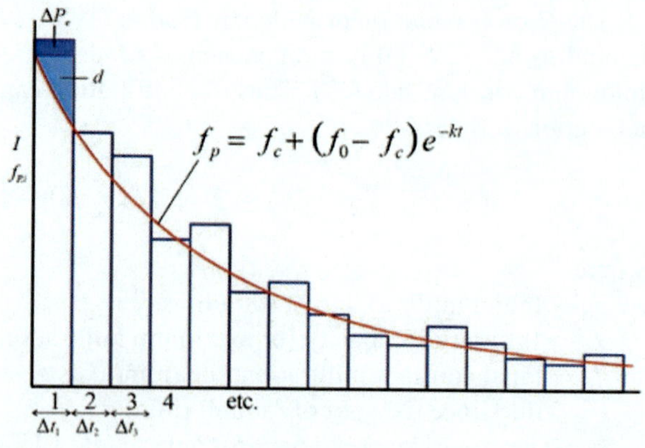

Fig. 2.11 Solving Horton's infiltration model

d = Contribution to depression storage capacity (mm). This will continue in subsequent time intervals until the depression storage maximum capacity d_{max} is reached.

$f_{p,i}$ = Infiltration capacity at the start of interval i (mm/h)

$f_{p,i+1}$ = Infiltration capacity at the end of interval i (mm/h)

I_i = Effective rainfall intensity during interval $[i, i+1]$ (mm/h)

Δt_i = Duration of time interval $[i, i+1]$ (mins) and terms in Horton's equation are as given in the text in the previous section.

At starting, the infiltration capacity is subjected to a finite value (f_0). As the time approaches infinity, the associated soil profile reaches a constant infiltration capacity (f_c) (Horton 1940; Hillel 1998). Horton's equation seems convenient as this provides a good fit for data than the Kostiakov equation. However, few limitations are involved, and these are:

- At the starting estimation f_0 is difficult.
- The infiltration rate would be over-predicted by assuming infiltration existence at f_p, the maximum rate, this ignores occasions during a storm when the actual rainfall rate is less than f_p.
- Deep percolation losses are ignored.

Constant loss rate method: the $\varphi - index$ represents the constant infiltration rate (mm/h), this approach works on the principle of higher water level, above the $\varphi - index$ horizontal line (i.e., the coloured part of Fig. 2.12a), equal to the water depth under the quick response hydrograph in Fig. 2.12b.

Here, 1, 2, …, M intervals of rainfall blocks present in the hyetograph, duration Δt_m (h) and intensity I_m (mm/h) as shown in Fig. 2.12a, it contributes to rapid response runoff. Then, the value of φ which satisfies the following relationship:

$$r_d = \sum_{m=1}^{M} (\Delta P_m - \varphi \Delta t_m) \tag{2.17}$$

where

r_d = Water depth under the quick response hydrograph

ΔP_m = Observed rainfall in block m = $I_m \times \Delta t_m$

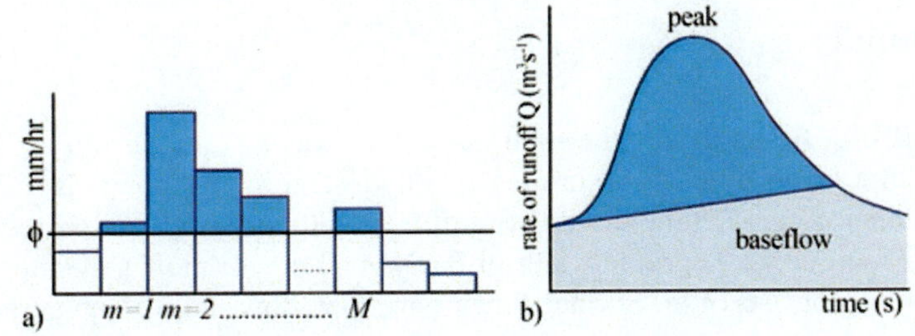

Fig. 2.12 **a** Rainfall hyetograph, **b** quick response runoff hydrograph

As the total rainfall block numbers contribute to r_d is unknown, to determine M (and therefore $\varphi - index$) should be carried out using Eq. 2.17 by trial and error. The $\varphi - index$ approach has the following limitations:

- The infiltration rate assumes as constant. Infiltration usually decreases with the increased storms and soil moisture content;
- This believes that infiltration always continues at the rate of $\varphi - index$ mm/hour regardless of the availability of sufficient rainfall;
- The approach requires a known quick response runoff hydrograph before computing infiltration, and this practice is unusual in most applied hydrological studies.

Overall, this approach is a straightforward method to estimate total infiltration but often fails to provide the temporal pattern of the infiltration.

Available infiltration models are of various categories, subject to the purpose of the model, boundary conditions, and the properties of subsurface systems. The selection of the model is always site-specific. The *Soil Conservation Service (SCS)* method commonly uses a semi-empirical infiltration model within the soil dynamics and hydrology study (USDA-SCS 1972). *Physically-based infiltration models* applied on homogeneous soil profiles. These models included the Green-Ampt method, Philip's infiltration equation, Burger infiltration equation, and Parlange. Green and Ampt (1911) model, an explicit approach, describe Darcy's infiltration law. Philip (1957, 1969) presented the first analytical solution to the Richards equation for vertical and horizontal infiltration. For horizontal infiltration, Philip estimates that the cumulative and instantaneous infiltration rates in terms of sorptivity. This is a function of initial and boundary water contents, i.e., $Sorptivity = f(q_o, q_i)$ during the elapsed time since water application. *The infiltration model for ponding conditions* tends to accumulate over the soil surface as the available rainfall exceeds the infiltration rate. In these conditions, the cumulative infiltration represents a function of soil properties, initial moisture content, and ponding depth. The *infiltration model for non-ponding conditions* describes the water supply rate to the soil surface as less than the soil infiltrability. This process depends on the water supply rate, initial soil moisture content, and K_{sat}. *The wetting and drying model* offers an unsteady diffusion of water into the soil through alternate infiltration and exfiltration.

2.6 Runoff

The runoff of a basin area is the total amount of water draining into a river or a reservoir in a given time. A basin area is a hydrologic unit known as a drainage basin, watershed, or catchment area. Runoff is broadly classified into *direct runoff* and percolation down to groundwater or *baseflow*. Direct runoff generates shortly after a rainfall consists of (a) Surface flow or overland flow and (b) Sub-surface flow or interflow.

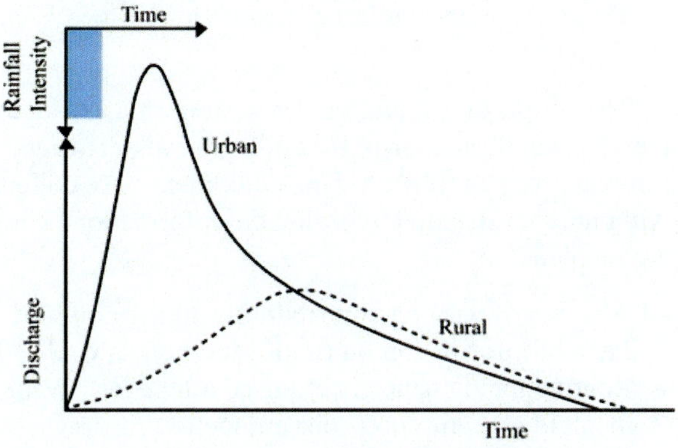

Fig. 2.13 Flood hydrographs for urban and natural drainage basins (Chow et al. 2013)

In an urban area lacking adequate stormwater management controls, increased imperviousness leads to a cascade of effects with increased overland flows. Rivers in highly urbanized areas are sometimes called "peaky" because of low flows under dry conditions and high flow (high volumes and high peak flows) during monsoon. As stated in Sect. 2.2, there can be a 3–5 folds increment in the surface runoff associated with the reduction in infiltration. A *hydrograph* is a graph based on the water flow versus time past a specific point or cross-section. For a given event, flood hydrographs, i.e., peak discharge over the duration, occur in urban versus rural or forested watersheds shown in Fig. 2.13.

The factors that influence the runoff are water flow from a drainage basin, hydrological, meteorological, and basin characteristics. The storm attributes consist of the type or nature of the storm and season, intensity, duration, areal distribution, storm movement direction, and precipitation. Meteorological characteristics include temperature, wind, humidity, and pressure variation. Basin characteristics comprise size, shape, topography, LULC, orientation, geology, and drainage density.

2.6.1 Runoff Data Collection

Usually, two types of equipment are involved in measuring runoff: (i) Volumetric equipment and (ii) Continuous or Through-flow equipment. Volumetric data collection includes simple tanks and complex tank systems; equipment has the following advantages:

- Suitable to measure minor runoff volumes;
- The essential equipment is a simple tank with a known proportion of the flow is collected, and the total flow can be computed by multiplying according to the proportions.

- This equipment could be manufactured locally and can be easy to replicate in experiments.

The simple tank equipment was subjected to a significant limitation with its physical size, depth, and top of the collection vessel needed to be installed at a lower elevation than the runoff area. This equipment only collects 'lump sum' runoff volumes without other detailed hydrological information. Thus, the following disadvantages are involved:

- Absence of varying contributions during complex storms;
- Lack of information on runoff duration as well as initial infiltration;
- Routine maintenance is required after every event and is often not suggested for the field station runoff measurements.
- The possibility of risks of over-filling of the tanks could lead to loss of accurate data.

On the other hand, periodic or continuous data collection methods could be depending on the physical properties of the flow and characteristics of the site. Continuous systems include two runoff measurements. These are: (i) Natural controls for runoff measurement: rating curves, velocity area method, streamflow networks, float, and chemical gauging; and (ii) Artificial controls for runoff measurement: flumes, weirs, ditches, and existing structures. Continuous systems offer the following advantages:

- Provide flow durations, peak flows, and starting of runoff with rainfall,
- Data recordings could cover several runoff events, and
- Equipment is well-suited for remote field sites.

This system is expensive similar to recording rain gauges and often requires complex machinery. Theoretically, runoff peak and volume could be estimated to aid the selection between two measuring methods.

2.6.2 Estimation of Runoff

Estimates of runoff are essential guides to determine the required size and capacity (peak flow and flow volume) of the measuring equipment and aid the design specifications of structures for rainwater harvesting, stormwater management, and groundwater recharges explained in respective chapters of this book. The runoff from rainfall could be estimated through:

a. Empirical formulae
b. Infiltration method
c. Rationale method
d. Unit hydrograph method
e. Water quality treatment volume
f. Water balance

g. Others

a. **Empirical formulae**

These apply to the region of origin, and cautions are required if the region's characteristics have been subjected to urbanization disturbances, viz. settlement, land use pattern changes. Additional parameters (i.e., third or fourth) need to be included for climatic or catchment characteristics along with rainfall-runoff relationships. Widely used empirical runoff estimation formulae are Binnie's Percentages, Barlow's Tables, Strange's Tables, Inglis and Desouza Formula, and Khosla's Formula.

b. **Infiltration method**

Infiltration is a complex process but one of the fundamental approaches to estimate runoff using any available infiltration models described in this section.

c. **Rationale method**

Flow estimation considers the entire drainage area as a single unit and outflow at the most downstream point only. The peak runoff from a storm of intensity I is given as:

$$Q_p = KCAI_p \tag{2.18}$$

where

Q_p = Peak discharge

A = Catchment area

C = Runoff coefficient

Usually, this value is region-specific; Table 2.6 is an example.

K = A conversion factor depends on the selected units. For example, if Qp is expressed in m^3/s, A is expressed in hectares and I in mm/h, then

$$K = \frac{1}{360}$$

I_p = Maximum rainfall intensity (mm/h) based on the return period and the '*time of concentration*' of the catchment, Tc. The Tc is the required duration for water to travel from the farthest upstream point of the catchment to the catchment's outfall. *Average Recurrence Interval* (ARI) data are the static precipitation/rainfall return periods, often available for regional standards of designing an event. The IDF analysis is widely applicable in this method.

Thus, Eq. (2.18) can be written as,

$$Q = \frac{CIA}{360}\ (\mathrm{m^3/s}) \tag{2.18a}$$

A composite runoff coefficient (C_w) should be applied for a catchment composed of multiple lands use by weighting C values as per the following equation:

Table 2.6 Runoff coefficients (C) for different LULC and return periods (recommended by the American Society of Civil Engineers and Water Environment Federation)

Description of area	Runoff coefficient (C) for the different return period			
	10 years	25 years	50 years	100 years
Business				
Downtown	0.70–0.95	0.77–1.05	0.84–1.14	0.88–1.19
Neighborhood	0.50–0.70	0.55–0.77	0.6–0.84	0.63–0.88
Residential				
Single-family	0.30–0.50	0.33–0.55	0.36–0.60	0.38–0.63
Multi-unit, detached	0.40–0.60	0.44–0.66	0.48–0.72	0.50–0.75
Multi-unit, attached	0.60–0.75	0.66–0.83	0.72–0.90	0.75–0.94
Residential (sub-urban)	0.25–0.40	0.27–0.44	0.30–0.48	0.31–0.50
Apartment	0.50–0.70	0.55–0.77	0.60–0.84	0.63–0.88
Industrial				
Light	0.50–0.80	0.55–0.88	0.60–0.96	0.63–1.00
Heavy	0.60–0.90	0.66–0.99	0.72–1.08	0.75–1.13
Parks, cemetries	0.10–0.25	0.11–0.28	0.12–0.30	0.13–0.31
Playgrounds	0.20–0.35	0.22–0.39	0.24–0.42	0.25–0.44
Railroad yard	0.20–0.35	0.22–0.39	0.24–0.42	0.25–0.44
Unimproved	0.10–0.30	0.11–0.33	0.12–0.36	0.13–0.38
Open surface				
Pavement				
Asphaltic and concrete	0.70–0.95	0.77–1.05	0.84–1.14	0.88–1.19
Brick	0.70–0.85	0.77–0.94	0.84–1.02	0.88–1.06
Roofs	0.75–0.95	0.83–1.05	0.90–1.14	0.94–1.19
Lawns, sandy soil				
Flat, 2%	0.05–0.10	0.06–0.11	0.06–0.12	0.06–0.13
Average, 2–7%	0.10–0.15	0.11–0.17	0.12–0.18	0.13–0.19
Steep, 7%	0.15–0.20	0.17–0.22	0.18–0.24	0.19–0.25
Lawns, heavy soil				
Flat, 2%	0.13–0.17	0.14–0.19	0.16–0.20	0.16–0.21
Average, 2–7%	0.18–0.22	0.20–0.24	0.22–0.26	0.23–0.28
Steep, 7%	0.15–0.35	0.17–0.39	0.18–0.42	0.19–0.44

$$C_w = \frac{\sum_{i=1}^{N}(C_i A_i)}{\sum_{i=1}^{N}(A_i)} \tag{2.19}$$

where

C_i = Runoff coefficient for individual land use in the catchment

A_i = Area of the individual land use in the catchment

N = Total number of land uses in the catchment.

The *rational method* is an approximate technique based on some assumptions:

- The rainfall distributed uniformly over the catchment;
- The duration of the catchment is at least equal to the T_c of the catchment. In other words, it will take rain to fall for at least T_c for all sections of the catchment to contribute flow to the catchment's outfall. Factors affecting T_c are the surface roughness, channel shape, and flow patterns, and slope. Thus, T_c is computed by summing up all the overland sheet flow and travel times (T_t) for the consecutive components of the drainage conveyance system i.e.

$$T_t = \frac{L}{V} \tag{2.20}$$

where

L = Overland sheet flow path length (m)

V = Average velocity (m/s)

$$= \frac{1}{n} R^{\frac{2}{3}} S^{\frac{1}{2}} \text{(using Manning's equation)} \tag{2.20a}$$

S = The slope of the surface (m/m)

R = Hydraulic radius (m)

n = Manning's roughness coefficients

Sheet flow passes over plane surfaces, i.e., parking lot, farm fields, and lawns. For the overland sheet flow path up to 50 m, the overland sheet flow travel time can be determined using the Kinematic Wave Equation (DPI, IMEA and BCC 1992):

$$T_t = \frac{6.94(\text{L.n})^{0.6}}{I^{0.4} S^{0.3}} \text{ min} \tag{2.21}$$

- The runoff coefficient is constant over the storm duration; and
- Absence of storage on the catchment during the storm.

Many regions could standardize the T_t based on the regional design events, and examples can be drawn from Australia. The Tt are 7 min and 6 min for the design events of the minor (10 years) and major (100 years), respectively (DPI, IMEA and BCC 1992).

Manning's Kinematics equation

$$T_t = \frac{5.48(nL)^{0.8}}{P_2^{0.5} S^{0.4}}$$

where

P_2 = The 2-years, 24-h rainfall (mm).

Example Problem 2.2 Estimate peak discharge for a drainage basin upstream of 81,000 m^2. The average overland slope is 2%. The lengths of the overland flow (AB), shallow concentrated flow over pavement (BC), open channel flow in the sewer pipe (CD), and available channel flow in vegetated swale (DE) are 25 m, 16 m, 305 m, and 122 m, respectively. Slope and Manning's roughness of the channel are 1.8% and 0.090, respectively. Land use for the drainage basin is:

- 80% area is a single-family occupied residential area.
- Graded/grass-covered area- silt loam soil to sandy soil, 3–20% slope.
- The overland flow area at the upper basin is of silt loam soil and a 2% slope. The runoff coefficient is 0.20.

Cross-sectional flow areas and the wetted perimeters for C-D and D-E are 0.123 m^2, 0.6 m^2, 1.07 m, and 3 m, respectively (Fig. 2.14).

Solution

Sheet flow (A-B)

Manning's roughness coefficient (for dense grass) = 0.025
Flow length = 25 m
2-year 24-h rainfall, P_2 = 75 mm
Land slope = 2%
As per Eq. 2.21

$$T_t = \frac{6.94(25 \times 0.025)^{0.6}}{75^{0.4}(0.02)^{0.3}} = 3\,\text{min}$$

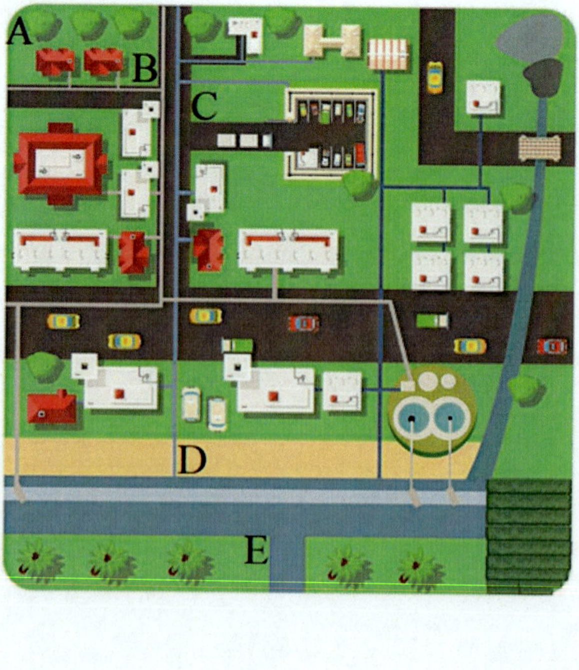

Fig. 2.14 Example problem 2.2

Shallow concentration flow (B-C)

Paved surface, with an average velocity of 0.9 m/s on the watercourse slope of 0.02. The travel time is as per Eq. 2.20:

$$T_t = \frac{L}{V} = \frac{16}{0.9 \times 60} = 0.3\ \text{min}$$

Open channel flow

	C-D	D-E (to the main channel)
Cross sectional flow area (m^2)	0.123	0.6
Wetted perimeter (m)	1.07	3
Hydraulic radius (m)	0.115	0.2
Manning's n	0.013	0.06
Velocity (m/s)	$\frac{1}{0.013}(0.115)^{\frac{2}{3}}(0.018)^{\frac{1}{2}}$ $= 2.44$	$\frac{1}{0.06}(0.2)^{\frac{2}{3}}0.016^{\frac{1}{2}}$ $= 0.72$
Flow length (m)	305	122
T_t (min)	2.03	2.82

Thus, T_c for the catchment is $(3 + 0.3 + 2.03 + 2.82)$ or $7.88 \approx 8$ min. A rainfall duration of 8 min engages to obtain the rainfall intensity:

- Rainfall depths for the 5-min duration for 25 and 50 years are 16 mm and 18 mm, respectively. Thus the intensities are 157.48 and 178 mm/h.
- The runoff coefficient for single residential area and grassed land are 0.40 and 0.25, respectively. Weighted runoff coefficients are 0.32 and 0.05, i.e., the total weighted runoff coefficient is 0.37.

  Peak runoff calculation

$$Q_{25} = 1.1 \times 0.37 \times \frac{157.48}{10000} \times 81000 = 519.2\ \text{m}^3/\text{s}$$

$$Q_{50} = 1.2 \times 0.37 \times \frac{178}{10000} \times 81000 = 640.1\ \text{m}^3/\text{s}$$

d. **Unit hydrograph method**

The unit hydrograph is a direct runoff hydrograph that results from 1 unit of continuous rainfall distributed uniformly over the catchment during a given time. In hydrograph analysis, the storm hyetograph (rainfall intensity versus time, input function) is converted to the direct runoff hydrograph (output) usual design practice using a unit hydrograph (transfer function) (Fig. 2.15), and the process is called *convolution*. Unit hydrograph method usually applies to moderate-size catchments areas of fewer than 5000 km^2.

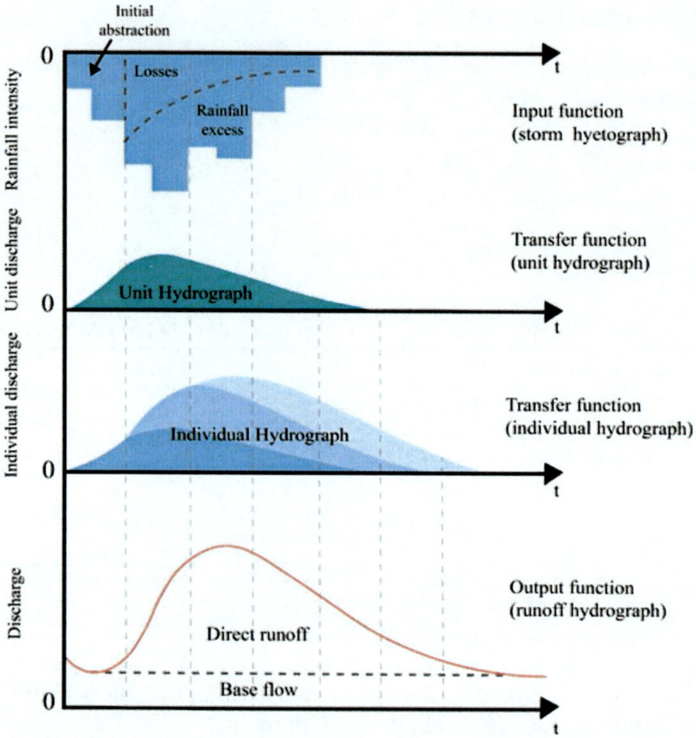

Fig. 2.15 Relationship among the storm, unit, and direct runoff hydrographs

US Soil Conservation Service Method or NRCS method is based on the rainfall/runoff relation for the triangular hydrograph. This method is applied to obtain the peak flow (q) of a known runoff event volume. Hence, peak flow, i.e., runoff rate (m^3/s), is defined by:

$$q = \frac{0.0021\,QA}{T_p} \tag{2.22}$$

where

Q = The area under the hydrograph, runoff volume (mm depth)
A = Area of the catchment, ha
T_p = Time to peak, hours

$$= \frac{\text{Excess rainfall duration}}{2} + T_L$$

T_L = Lag time is an approximation of the mean travel time
= 0.6 × Time of concentrationthe mean travel time (T_c).

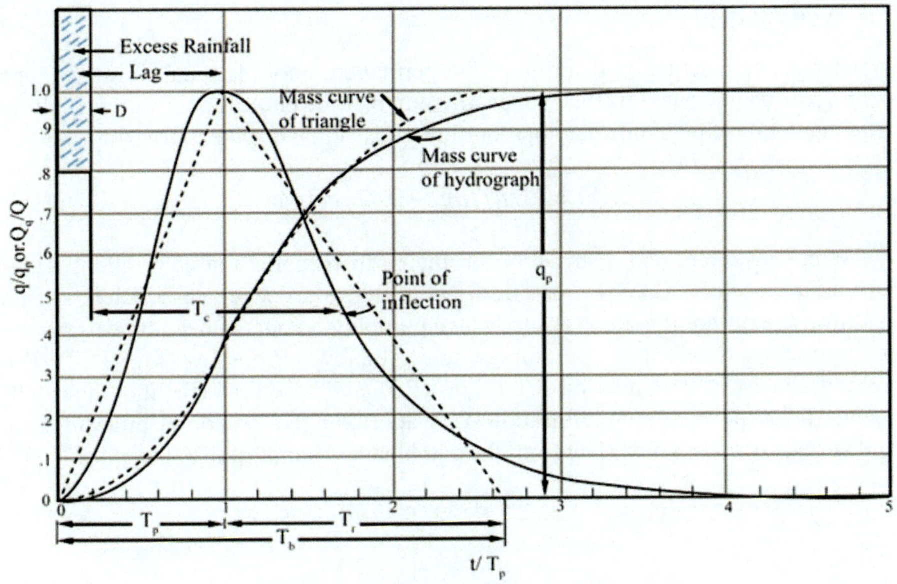

Fig. 2.16 Dimensionless unit hydrograph by NRCS (1972)

The NRCS dimensionless unit hydrograph (NRCS 1972) (Fig 2.16) is practically used and recommended to design conservation practices or the BMPs. All discharge ordinates and time values are divided by the peak discharge and the time to peak, respectively. The average of these dimensionless unit hydrographs (UH) is then computed as:

The time-base (T_b) of the dimensionless UH $= 5 \times T_p$

$= 3/8$ of the total volume occurred before the time to peak.

The *inflection point* on the falling limb exists at about 1.7 times the time to peak, and the hydrograph has a curvilinear shape. This curvilinear hydrograph is generally approximated by a triangular UH with similar characteristics.

e. **Water Quality Volume**

The Water Quality Volume (Q_{wv}) is required to remove a significant stormwater pollution load. The Q_{wv} is calculated by the 85th percentile annual rainfall event, volumetric runoff coefficient (R_v), and the catchment area (A):

$$Q_{wv} = 85th\ percentile\ annual\ rainfall\ event \times R_v \times A \tag{2.23}$$

where

$R_v =$ Volumetric runoff coefficient

$= 0.05 + 0.009(LULC)$

LULC = Percent of impervious cover (%)

f. **Water Balance**

Water balance calculation determines the permanent pool of water during average or extreme conditions. It is the change in volume of the endless pool water resulting from the total inflow minus the total outflow (i.e., actual or potential):

$$\Delta V = \Sigma Inflow - \Sigma Outflow \tag{2.24}$$

Water movement and distribution on the earth comprises stream inflow, tributary inflow, surface water diversion, surface water return, deep percolation, stream outflow, stream aquifer flux, groundwater pumping, groundwater inflow, groundwater storage, transit loss, soil storage, and groundwater outflow (Fig. 2.17). The inflows consist of Precipitation (P), runoff (Ro), and base flow (B_f) into the catchment. The outflows refer to infiltration (f), evaporation (E), evapotranspiration (ET), and surface overland flow (Of) from the catchment. Then, Eq. 2.24 becomes:

$$\Delta V = P + Ro + Bf - f - E - ET - Of \tag{2.25}$$

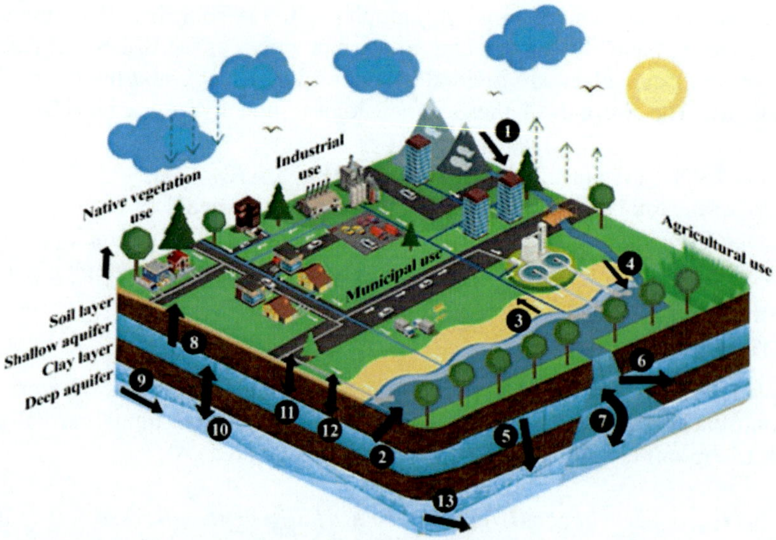

1. stream inflow, 2. tributary inflow; 3. surface water diversion, 4. surface water return, 5. deep percolation, 6. stream outflow, 7. stream aquifer flux, 8. groundwater pumping, 9. groundwater inflow, 10. groundwater storage, 11. transit loss 12. soil storage, 13. groundwater outflow

Fig. 2.17 Water balance

- Other methods

a. **Cook's Method**

The NRCS develops this more straightforward and generalized approach method similar to the Rational Method to estimate direct runoff (Q) from 24-h or 1 Day storm rainfall, and the required equation is:

$$Q = \frac{(P - Ia)^2}{(P - Ia) + S_{max}} \tag{2.26}$$

where

$$S_{\max} = \text{Potential maximum soil retention}$$
$$= \frac{1000}{CN} - 10$$

$$\text{Ia} = \text{Initial abstraction including surface storage, interception, evaporation, and infiltration before runoff initiated}$$
$$= 0.2S_{max}$$

The runoff coefficient has been selected from generic lists based on the LULC and infiltration capacity of the drainage surface. The estimated roof runoff coefficients are 0.7 to 0.95 for relatively frequent storms (Guisasola et al. 2011). Thus, it is urgent to develop a list of specific runoff coefficients for various roof types under diverse environmental, climatic conditions in the context of rainwater harvesting, described in Chap. 3.

b. **UK Transport and Road Research Laboratory (TRRL) Model**

This model is designed to overcome the associated data handling issues in many developing countries by developing significant data series rainfall-runoff correlations.

2.7 Groundwater

Groundwater is fresh water from precipitation that soaks into the soil and deposits among pore spaces, fractures, and joints within rocks and other geological formations. *Porosity* is the measure of areas within a material that can hold water under the ground, and *permeability* describes the distribution of pores based on their shapes. Groundwater is present in the saturated soil and rock below the water table. Rapid groundwater water movements occur for the shallow and permeable aquifer, and then wells are drilled to withdraw water. Groundwater tables might fluctuate over time due to changes in climate, streamflow, geologic changes, increase in impervious surfaces, and land-use changes. The groundwater in the *zone of aeration* or unsaturated area or vadose zone or zone of suspended water may enter through (i) infiltration of the surface water from rain, stream/reach/river, and wastewater; and (ii) capillary effect of the water from the saturation zone below the water table. Below this layer is the *zone of saturation* filled with water, and the water table is the boundary between these two layers.

2.7.1 Aquifer Formation

Aquifer refers to saturated rocks or geological formations to ensure permeability into wells and springs. Unconsolidated materials, i.e., gravel, quartzite, sand, and even silt, fit in suitable aquifers if they are well fractured. An *aquiclude*, i.e., clays, shales, etc., absorbs water but lacks significant water transmission. An *aquitard* also restricts transmission of a considerable amount of water, i.e., clay lenses interbedded with sand, a till, or a poorly fractured igneous or metamorphic rock. With the absence of interconnected pores, an *aquifuge* body is reluctant to absorb and transmit, viz. basalts, granites, etc.

Unconfined aquifer exposes directly to the ground surface. A *confined aquifer* with lower permeability exists between the aquifer and the ground surface, and the *aquitard* separates the ground surface and the aquifer as the confining layer (Fig. 2.18).

In the confined aquifer, water level exists under hydrostatic pressure. Wells within the confined aquifer contain higher water levels over the water-bearing formation until the local hydrostatic pressure within the well turns similar to the atmospheric pressure. These wells might be or not be flowing wells (Fig. 2.18). In contrast, water exists under atmospheric pressure in an unconfined aquifer, and wells drilled in

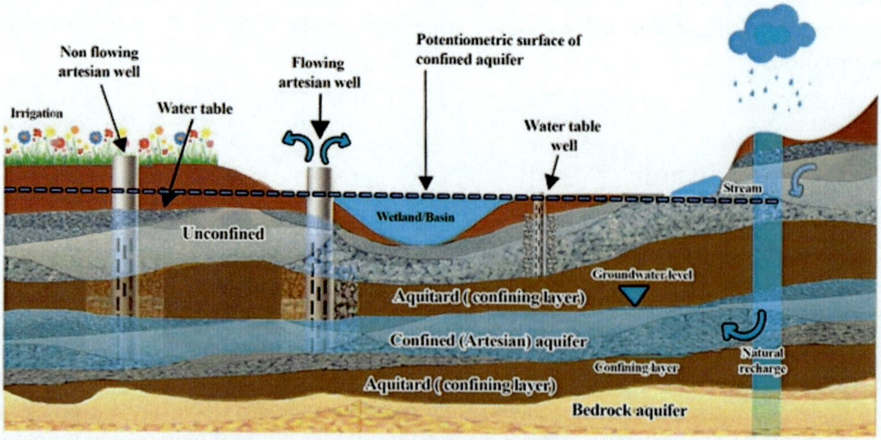

Fig. 2.18 Aquifer formation

aquifers having the local water table. A water table map or *potentiometric surface map* presents the wells' spatially distributed water levels within confined and unconfined aquifers.

Specific yield (S_y) is the volumetric fraction of the bulk aquifer volume is allowed to drain out of the aquifer due to gravity forces. On the other hand, *Specific Retention* (S_r) is the volume of water retained per unit volume of the aquifer after the specific yield. Generally, S_y will range from 0.01 to 0.45, depending on the aquifer material (Chin 2000):

$$Porosity(\varphi) = Specific\ yield(S_y) + Specific\ retention(S_r) \tag{2.27}$$

Transmissivity, T_t, is the flow per unit width of the aquifer under a unit hydraulic gradient. Then,

$$T_t = Kb \tag{2.28}$$

where K is the hydraulic conductivity and b is the saturated thickness of the aquifer.

Storage Coefficient or storativity is the water flow discharged from a confined aquifer per unit area per unit drop in the piezometric head. Therefore, the *storativity* is dimensionless, magnitudes ranging from 10^{-6} to 10^{-2} (Zekâi 1989). S*torativity* is generally lower than the corresponding values of the specific yield for unconfined aquifers (usually, by a factor of between 1000 to 10,000). Thus, change in the piezometric surface in a confined aquifer is significant than the unconfined aquifer. The storativity-width ratio is termed as the *specific storage* (Ss), i.e., the amount of water released from the storage over per unit porous medium recorded due to per unit declination in the piezometric head.

2.7.2 Groundwater Movement

Groundwater flow is almost laminar. Darcy's (1856) law states the water flow in saturated sand and found that the flow velocity was proportional to the hydraulic gradient. Figure 2.19 shows a vertical cross-section over a saturated porous media; two piezometers are located at a distance L apart with z, the height above a datum, and h, the height of water in the piezometer. Thus, the total piezometric heads are ($h_1 + z_1$) and ($h_2 + z_2$) at locations 1 and 2, respectively, then Darcy's law can be expressed mathematically as

$$\frac{Q}{A} = -K\frac{(h_2 + z_2) - (h_1 + z_1)}{L} = -Ki \tag{2.29}$$

Here,

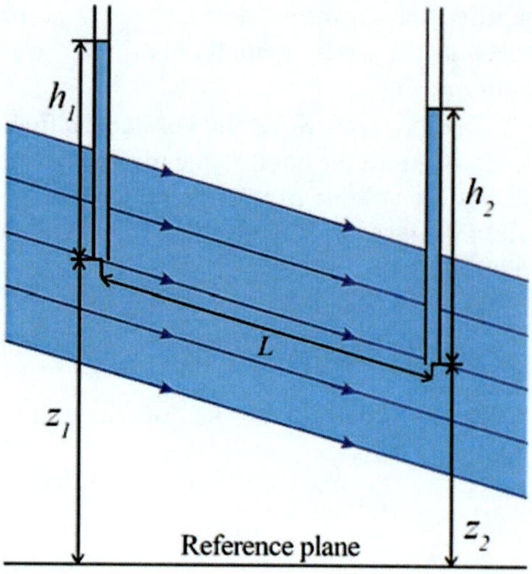

Fig. 2.19 Vertical cross-section of groundwater flow showing two piezometers 1 and 2 and their respective heads ($h_1 + z_1$) and ($h_2 + z_2$)

Q is the flow rate, A is the cross-sectional area of the sample perpendicular to the flow, i is the hydraulic gradient, and K is the constant of proportionality or saturated hydraulic conductivity or the *coefficient of permeability*.

The ratio (Q/A), which has the unit of velocity, is often termed the *Darcy velocity* or *specific discharge* or the *filtration velocity* and is different from the actual velocity through the porous media. The *actual velocity*, V_{act}, also termed the *seepage velocity*, is related to the Darcy velocity through

$$V_{act} = \frac{v}{\varphi} \tag{2.30}$$

where

v is the Darcy velocity, and φ is the porosity of the media.

Example Problem 2.3 During groundwater investigation in the basin stated in Example problem 2.1, the recorded details were as below:

Recharge area identified	70% of the total catchment area (i.e. 81,000 km^2) = 56,700 km^2
Annual precipitation	3000 mm
Infiltration	20% of precipitation
Transmissibility of the aquifer (from the pump tests in the discharge area)	6.57×10^3 lpd/m
Width of the aquifer	20 km
Hydraulic gradient (towards the discharge area from observation wells)	1.14 m/km

Determine whether all the pumpage comes from the recharge area?

Solution

$$\text{Annual recharge} = \left(56700 \times 10^6\right) \times \frac{20}{100} \times 3.00 = 34,020 \times 10^6\,\text{m}^3$$

$$\text{Pumpage, } Q = Tib = \left(6.57 \times 10^3\right) \times \frac{1.14}{1000} \times 20,000 = 149.8 \times 10^3\,\text{m}^3$$

$$Annual\ Pumpage = \left(149.8 \times 10^3\right) \times 365 = 54.7 \times 10^6\,\text{m}^3$$

Thus, with the present urbanized/impervious land use consideration, the recharge area seems sufficient for the entire pumpage.

2.7.3 Groundwater Recharge

Infiltration within the hydrological cycle contributes to groundwater storage through deep percolation by natural recharge. The percolating water can be from rainfall or water applied during irrigation of fields. However, continuous withdrawal might deplete the groundwater reservoir towards empty. Water smart city refers to groundwater recharge to replenish groundwater within the urban area. Groundwater recharge ensures:

- Reduce groundwater mining and possible saltwater intrusion in coastal regions.
- Lessen pumping cost due to the filled water table or the piezometric surface.
- The safe yield of an aquifer is the amount of water it receives as recharge. Consequently, low recharge means a low yield of the aquifer.
- Prolonged dry spells in rivers fed by groundwater—the baseflow in most rivers are derived from groundwater contribution. Thus, the water table experiences a drastic lowering underneath the river bed; receiving baseflow assistance from such aquifers would mean that flows in the associated rivers can change from perennial to intermittent.

Three techniques are practicing to recharge groundwater, i.e., natural recharge, bank filtration, and artificial recharge (Fig. 2.20). *Natural recharge* is the infiltration of direct precipitation. *Bank filtration* is also a natural recharge through a well placed near a stream. This well can draw streamflow into an aquifer, thereby enhancing the natural recharge to the aquifer. In artificial recharge, streamflow is directed or pumped to an artificial recharge basin, where water percolates downward to the underlying aquifer.

Similar to insufficient recharge, the excess recharge is also undesirable due to the following reasons:

- Increased risk of groundwater contamination by surface contaminants
- Reduction in bearing capacity of foundation materials
- Moisture ingress into sub-surface structures, e.g., pipelines, tunnels

1: Natural recharge, 2: Bank filtration, and 3: Artificial recharge

Fig. 2.20 Groundwater recharge techniques

- Risk of flotation of shallow-founded pipes and subsurface services
- General difficulty with repair, maintenance, and new construction of sub-surface structures.

Unconfined aquifers with sufficient rainfall and the reasonably previous ground surface are recharged naturally with significant rain infiltration. On the contrary, confined aquifers face challenges to restore naturally because of their overlying aquicludes that prevent any vertical movement of any infiltrated water into the aquifer. However, confined aquifers might naturally be restored when the overlying formation is leaky. Still, even in such situations, the recharge amount is unlikely to be sufficient to replenish the water withdrawn from such aquifers and requires an artificial recharge. Details on artificial recharge have explained in Chap. 6.

(a) **Flow in Unconfined Aquifers**

In unconfined aquifers, the water table found in the upper boundary of the groundwater flow region and direct analytical solution of the Laplace equation fails to determine this flow. The saturated thickness of the aquifers drops in the flow direction (Fig. 2.21). Due to the absence of recharge or evaporation, the amount of flowing water upstream and downstream remains similar.

In Darcy's law, as the downstream cross-sectional is smaller, the hydraulic gradient is higher on this side relatively constant in unconfined flow, and it increases in the flow direction. To overcome this issue, Dupuit (1863) adopted certain simplifying assumptions, and these are:

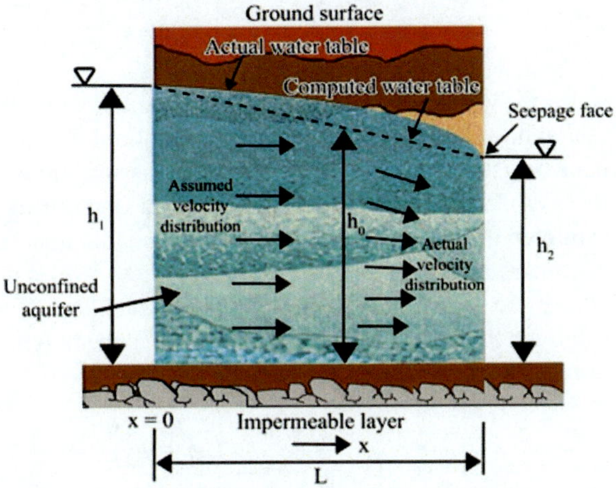

Fig. 2.21 Steady flow within an unconfined aquifer (modified from Todd 1980)

i. In an unconfined flow system, the hydraulic gradient is the slope of the water table, and
ii. The streamlines pass horizontally, and the equipotential lines are vertically within a lower hydraulic gradient.

For steady flow in an unconfined without recharge or evapotranspiration by applying Dupuit assumptions on Darcy's law at any vertical section, the groundwater flow per unit width of the aquifer (q):

$$q = -Kh\frac{dh}{dx} \tag{2.31}$$

Here,
h = Saturated thickness of the unconfined aquifer
$\frac{dh}{dx}$ = Hydraulic gradient.
In Fig. 2.21, L = flow length, h_1 = head at the origin i.e. x = 0, and h_2 = head at a distance L i.e. x = L. Applying boundary conditions, i.e., at x = 0, h = h_1; at x = L, h = h_2 (Fig. 2.21), Eq. 2.31 can be written as:

$$\int_0^L q dx = -K \int_{h_1}^{h_2} h dh \tag{2.32}$$

Then,

$$q = \frac{K}{2L}\left(h_1^2 - h_2^2\right) \tag{2.33}$$

As per Dupuit equation (i.e. Eq. 2.33) the water table is in parabolic form. The capillary zone has been ignored, the water table slope at the upstream boundary of the aquifer is as:

$$\frac{dh}{dx} = -\frac{q}{Kh_1} \tag{2.34}$$

As the water body contains constant fluid potentialities hence, $h = h_1$ is an equipotential line. Accordingly, the water table should be horizontal within this section; this is inconsistent with Eq. (2.34). Dupuit's assumptions of horizontal flows can explain this. While the actual velocities act as a downward vertical component, a greater saturated thickness (i.e., the higher water table than the aquifer base) is required for a similar discharge.

For the *steady unconfined flow with recharge or evapotranspiration* on known saturated thickness (as shown in Fig. 2.22), the height of the water table between two points located on L distance is:

$$h(x) = \left[h_1^2 - \frac{(h_1^2 - h_2^2)x}{L} + \frac{R}{K}(L - x)x \right]^{0.5} \tag{2.35}$$

where

h (x) = hydraulic head at a distance x from the origin

h_1 = head at the origin

h_2 = head at the distance L

and R = recharge rate.

If the evapotranspiration (ET) is higher than the recharge (R_i), then R_i should be replaced by ET using a negative sign (i.e., −ET) in Eq. 2.35. In the absence of recharge or evapotranspiration, Eq. (2.35) will be:

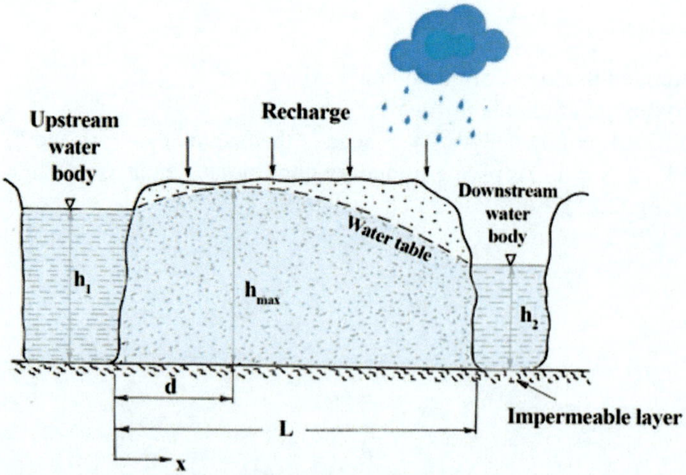

Fig. 2.22 Recharge held in steady unconfined flow (modified from Fetter 2001)

$$h(x) = \left[h_1^2 - \frac{(h_1^2 - h_2^2)x}{L}\right]^{0.5} \tag{2.36}$$

Equation (2.36) is known as Dupuit parabola. Now, by differentiating Eq. 2.35 and considering $q_x = -Kh\frac{dh}{dx}$, the discharge per unit width at any section is given by:

$$q(x) = \frac{K(h_1^2 - h_2^2)}{2L} - R_i\left(\frac{L}{2} - x\right) \tag{2.37}$$

If recharge (R_i) continues, a *groundwater divide* with a crest in the water table persists. If d is the distance from the origin to *groundwater divide*, then for q (x) = 0 and x = d into Eq. (2.37) yields:

$$d = \frac{L}{2} - \frac{K}{R_i} \times \frac{(h_1^2 - h_2^2)}{2L} \tag{2.38}$$

At the groundwater divide, the maximum water-table height (i.e., h_{max}) from the aquifer base is by substituting x with d in Eq. 2.35 as:

$$h_{max} = \left[h_1^2 - \frac{(h_1^2 - h_2^2)d}{L} + \frac{R}{K}(L - d)d\right]^{0.5} \tag{2.39}$$

(b) **Flow in confined aquifers**

Darcy's law directly applies to a steady groundwater flow in a homogeneous and isotropic confined aquifer of uniform thickness; a linear gradient or slope exists to the piezometric surface. Two observation wells/piezometers are available within the L distance (Fig. 2.23); using Darcy's law, the groundwater flow quantity is:

$$q = Kb\frac{dh}{dx} \tag{2.40}$$

The hydraulic head (h) at some intermediate distance, x between Piezometer 1 and Piezometer 2 as:

$$h(x) = h_1 - \frac{q}{Kb}x \tag{2.41}$$

(c) **Flow into horizontal galleries dug down to the impervious soil layer**

With a depth of groundwater above impervious soil layer (H) and depth of water table in the gallery (h_1), if the parabolic phreatic surface drops from H to h in the distance L. Here, the horizontal distance for the water depth (h) is x from the face of the gallery. The quantity of water flowing into the trench from one side per unit length of the shoreline is (Fig. 2.24):

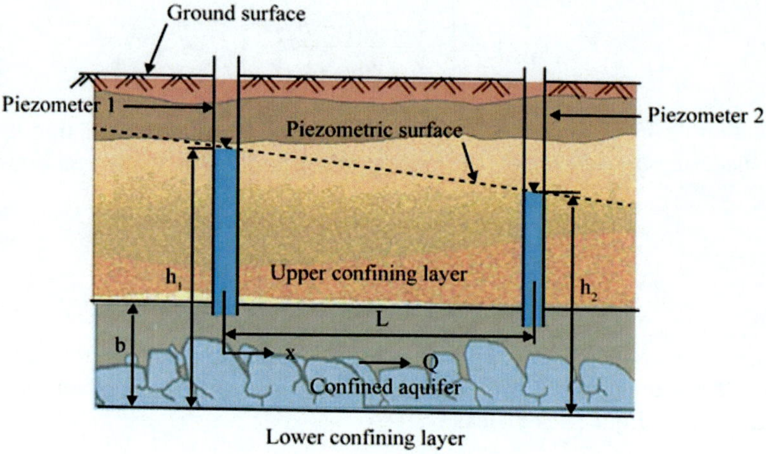

Fig. 2.23 Steady flow within a confined aquifer (modified from Fetter 2001)

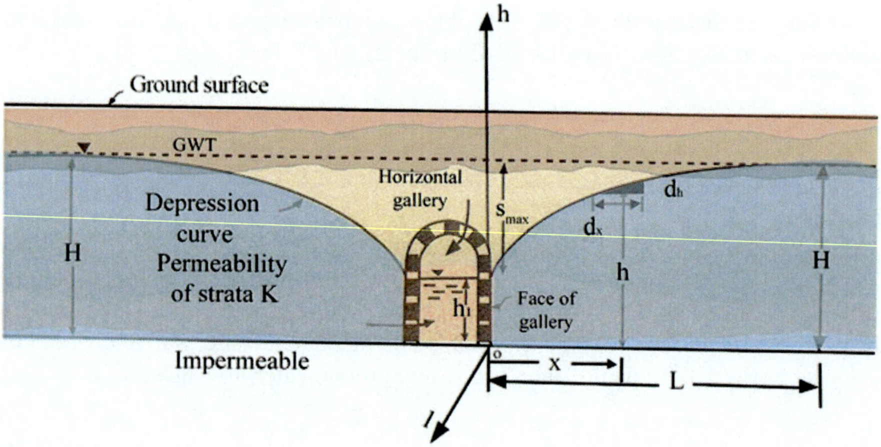

Fig. 2.24 Flow into horizontal galleries

$$q = VA = Ki(h \times 1) = Kh\frac{dh}{dx} \tag{2.42}$$

Integrating,

$$qx = \frac{Kh^2}{2} + c_1$$

When,

$$x = 0, h = h_1; c_1 = -\frac{Kh_1^2}{2}$$

$$q = \frac{K}{2x}\left(h^2 - h_1^2\right) \tag{2.43}$$

Equation 2.43 is parabolic. Replacing h = H when x = L

$$q = \frac{K}{2L}\left(H^2 - h_1^2\right) \tag{2.44}$$

From Eqs. (2.43) and (2.44), the equation to the phreatic line is:

$$h = \sqrt{h_1^2 + \frac{x}{L}\left(H^2 - h_1^2\right)} \tag{2.45}$$

The quantity of water flowing into the gallery of length l from both sides

$$Q = 2ql = \frac{Kl}{L}\left(H^2 - h_1^2\right) \tag{2.46}$$

If the water table depleted in the gallery, i.e., h_1 decreases, L increases upon increasing h_1, L decreases.

Land use, geology, and soils are significant concerns in the site selections for water spreading or infiltration recharging systems. There are two considerations: water needs to move throughthe vadose zone and ensures water moving through the aquifer is away from infiltration recharge sites to raise water build-up of groundwater mound or ridge. Therefore, for the groundwater recharge system, the selected area should meet the following characteristics:

- The ground surface should be permeable to conduct infiltration;
- The 'Vadose' zone must be permeable and free from clay layers;
- The deserving aquifer for recharge should be unconfined and permeable with sufficiently thick to avoid the rise of groundwater mounds adjacent to the land surface;
- The groundwater table needs to be deep and possibly maintained 8–10 m below the ground surface.

Example Problem 2.4 From the pumping tests of a semiconfined aquifer of thickness 30 m and permeability 20 m/d, the estimated recharge rate from an unconfined aquifer over an aquitard of 2.1 m thick is 40 mm/year. The average piezometric surface in the semiconfined aquifer is 16 m below the water. Determine (a) hydraulic characteristics of the aquitard and the aquifer; (b) Within this aquifer, if a well expect to pump 4000 m^3/day, determine the required area for recharge?

Solution

Recharge through *aquitard* per unit area (1 m^2), if K' is the hydraulic conductivity for the aquitard.

$$Q = K' i A$$

$$\frac{0.040}{365} = K' \times \frac{16}{2}(1 \times 1)$$

$$\text{Then, } K' = 1.37 \times 10^{-5}\ \text{m/day}$$

$$\text{Leakance} = \frac{K'}{b'} = \frac{1.37 \times 10^{-5}}{2.1} = 6.52 \times 10^{-6}\ \text{day}^{-1}$$

Hydraulic resistance is the reciprocal of Leakance,

$$c = \frac{b'}{K'} = 1.53 \times 10^{5}\ \text{days}$$

Hydraulic characteristics of the *aquifer*,

$$T = Kb = 20 \times 30 = 600\ \text{m}^2/\text{day}$$

Leakage factor,

$$B = \sqrt{Tc} = \sqrt{600 \times \left(1.53 \times 10^{5}\right)} = 9590\,\text{m}$$

To produce pumpage of 4000 m^3/day, the required recharge area is:

$$\frac{0.040}{365} A = 4000;\ A = 3.65 \times 10^{7}\ \text{m}^2 = 36.5\ \ \text{km}^2$$

2.8 Water Smart City and Water Quality

Generally, rainwater is relatively free from impurities in the hydrological cycle except those added naturally during the precipitation generation process within the atmosphere. Rainwater is slightly acidic and contains few dissolved minerals. According to WHO (2020), chemical concentrations in rainwater are generally within acceptable limits. Usually, rainwater lacks minerals and often fails to attract communities with the available mineral-rich natural waters. The different phases of the system influence the water quality for a rainwater harvesting system, described in Chap. 3.

Increased impervious surfaces pose adverse impacts on the urban hydrological cycle. Thus, reductions in infiltration limit the recharge of groundwater resources and generate higher runoff peaks in a shorter duration. Acidic rainfall may result in dense areas with higher industrial growth and also the presence of tall smokestacks. Rainwater quality for a rainwater harvesting system needed for improved water conservation, water supply during extreme conditions, minimized water costs for the end-user, improved groundwater recharge, etc., depends on harvesting techniques and storage and consumers choice, described in Chaps. 4, 5, and 6. Overall, based on the consumption nature, political and social settings, the appropriate standards for rainwater quality would be applicable.

References

Chin DA (2000) Water-resources engineering. Prentice-Hall Inc., Upper Saddle River, New Jersey.

Chow VT (2009) Open-channel hydraulics. The Blackburn Press, Caldwell, N.J.

Chow VT, Maidment DR, Mays LW (2013) Applied hydrology. McGraw-Hill Professional; McGraw-Hill [distributor], New York; London.

Critchley W, Siegert K (2013) Water harvesting: a manual for the design and construction of water harvesting schemes for plant production. Scientific Publishers, Jaipur (India).

Darcy H (1856) Les Fontaines Publiques de la Ville de Dijon [The public fountains of the city of Dijon]. Dalmont, Paris.

DPI, IMEA and BCC (1992) Queensland Urban Drainage Manual (QUDM), prepared by Neville Jones & Associates and Australian Water Engineering for DPI, IMEA & BCC, Brisbane.

Dupuit J (1863) Études Théoriques et Pratiques sur le Mouvement des Eaux dans les Canaux Découverts et a Travers les Terrains Perméables (Theoretical and practical studies on water movement in open channels and across permeable terrains), deuxième édition, Dunod, Paris, pp 229–293.

Fetter CW (2001) Applied hydrogeology. Pearson Education, Upper Saddle River, NJ.

Green WH, Ampt GA (1911) Studies on soil physics, Part 1, The flow of air and water through soils. J Agric Sci 4:11–24.

Guisasola A et al (2011) Roof selection for rainwater harvesting: quantity and quality assessments in Spain, 5. https://doi.org/10.1016/j.watres.2011.03.036

Hillel D (1998) Environmental soil physics. Academic Press Inc, New York, USA. https://www.elsevier.com/books/environmental-soil-physics/hillel/978-0-12-348525-0

Hillel D (2013) Introduction to soil physics. Academic Press.

Horton RE (1939) Analysis of runoff plot experiments with varying infiltration capacity. Trans Am Geophys. Union, Part IV.

Horton RE (1940) An approach towards a physical interpretation of infiltration capacity. Soil Sci Soc Am 5.

Johnson AI (1963) A field-method for measurement of infiltration. U.S. Gov. Print. Off, Washington, DC.

Johnson JH, James H (1971) Urban geology—an introductory analysis. Pergamon Press, Oxford.

Kale N et al (2019) 2019 water year report. Summary report and analysis for water year 2019: including atmospheric, streamflow, groundwater and lake level data – Thurston county water planning.

Kostiakov NA (1932) On the dynamics of the coefficient of water-percolation in soils and on the necessity of studying it from a dynamic point of view for purposes of amelioration. In: Transactions on 6th congress international soil science, Russian Part A, pp 17–21. http://ci.nii.ac.jp/naid/10011005232/en/. Accessed 9 April 2021.
Leopold LB (1968) Hydrology for urban land planning—a guide on the hydrologic effects of urban land use, U S geological survey Circ. 554, 18 pp.
Lewis MR (1937) The rate of infiltration of water in irrigation-practice. Eos Trans Am Geophys Union (Wiley) 18(2):361–368. https://doi.org/10.1029/TR018i002p00361
Linsley RK et al (1992) Water resources engineering. McGraw-Hill, New York.
Lull HW (1959) Soil compaction on forest and range lands. Forest Service, U.S. Dept. of Agriculture, Washington, DC.
Niemczynowicz J (1999) Urban hydrology and water management—present and future challenges. Urban Water 1(1):1–14. https://doi.org/10.1016/S1462-0758(99)00009-6
NRCS (1972) National engineering handbook, Section 4, Hydrology. U.S. Department of Agriculture, Washington, DC.
Philip JR (1957) The theory of infiltration: 1. The infiltration equation and its solution. Soil Sci 83:345–357.
Philip JR (1969) Theory of infiltration. In: Chow WT (ed) Advances in hydroscience. Elsevier, pp 215–296. https://doi.org/10.1016/B978-1-4831-9936-8.50010-6
Reed D (1999) Flood estimation handbook, volume 1: Overview. Institute of Hydrology, Wallingford.
Shaw EM et al (2011) Hydrology in practice, 4th edn.
Sherman CW (1931) Frequency and intensity of excessive rainfalls at Boston, Massachusetts. Trans Am Soc Civ Eng Am Soc Civil Eng 95(1):951–960. https://doi.org/10.1061/TACEAT.0004286
Todd DK (1980) Groundwater hydrology. Wiley, New York.
USDA (1957) Rainfall intensity-frequency regime. Technical Paper No. 29.
USDA (1961) Rainfall frequency atlas of the United States for durations from 30 minutes to 24 hours and return periods from 1 to 100 years. Technical Paper No. 40.
USDA-SCS (1972) National engineering handbook, hydrology Section 4. USDA, Washington, DC.
USEPA (2007) Better assessment science integrating point and nonpoint sources. Modeling Framework. National Exposure Research Laboratory, RTP, North Carolina.
WHO (2020) Water sanitation health. https://www.who.int
Wilson EM (1983) Engineering hydrology (Macmillan civil engineering hydraulics), 4th edn. Palgrave Macmillan. ISBN-10: 0333517172.
Wisler CO, Brater EF (1959) Hydrology. Chapman & Hall, Wiley.
WMO (1986) Manual for estimation of probable maximum precipitation, 2nd edn. Operational Hydrology Report No. 1, WMO—No. 332, Geneva.
Zekâi S (1989) Nonlinear flow toward wells. J Hydraul Eng Am Soc Civil Eng 115(2):193–209. https://doi.org/10.1061/(ASCE)0733-9429(1989)115:2(193)

Chapter 3
Rainwater Harvesting System

3.1 Introduction

A typical rainwater harvesting system comprises four main components are the catchment, the conveyance, the storage, and the distribution system in the building. This chapter described these components' details, design, and instalment consideration following the available codes and standards, management strategies, harvested water quality, and economic assessment. These components are further divided into smaller divisions considering the size and nature of the catchment surface, the designed use for the harvested water, the conveyance system, the type and position of water storage, and the properties of the material used for the construction of individual subdivisions. To facilitate architectural conceptualization, computation of 'roof footprint' has been described for rooftop catchment areas considering different roof shapes. Rainwater storage/tanks components, materials, position, and harvested water treatment modes for the pre-storage and post-storage have been described. Calculations of the storage performance evaluation and the efficiencies are also included. The distribution system includes pumps, pressurizing flow to end-use, bypass, and makeup water. Management strategies for these components are included. Workout examples are provided to develop componentwise understanding and design of a complete system with the necessary water treatment.

3.2 Catchment

A catchment is an exposed surface area to collect precipitation, and the surface runoff would eventually flow to a draining system or into a groundwater system. The catchment area for harvesting depends on rainfall, watershed slope, types of soil and vegetation, and the evapotranspiration ratio. However, catchment areas need to be heavily protected and pollution-free. For example, about half of Singapore's land area practices rainwater collection (Khoo 2009). The catchment size, shape,

© The Author(s), under exclusive license to Springer Nature Switzerland AG 2022

A. Akter, *Rainwater Harvesting—Building a Water Smart City*, Springer Water,
https://doi.org/10.1007/978-3-030-94643-2_3

degree of inclination, materials, and rainfall intensity-duration -frequency influence the harvested water quantity.

Any pavement area in a building could contribute to the collection system, including roof, verandas, balconies, sunshades, corners, car porch, and part of side-walls. Worldwide recommended catchment size varies due to demand and available rainfall intensity-duration-frequency. For instance, as a catchment area, the general building code recommended roof size for adopting rainwater harvesting in Australia is 50 m^2 (Hofstein et al. 2004; AS/NZS 2018; ABCB 2019a) compared to 300 m^2 in Bangladesh (BNBC 2020). Around India, rainwater harvesting is obligatory for the forthcoming building with a rooftop area of or above 100 m^2 (New Delhi), 222 m^2 (Bangalore), 250 m^2 (Madhya Pradesh), 300 m^2 (Andhra Pradesh), 500 m^2 (Rajasthan), 1000 m^2 (Himachal Pradesh, Uttar Pradesh, Mumbai) and 1,500 m^2 (Ahmedabad). Similarly, calculations for catchment areas are also varying country to country (Table 3.1). There is a need for footprint area rather than the surface area of the roof for catchment calculation. *The roof footprint* indicates the roof's plane projection area, or the area calculated from bird view is the footprint area. Roof footprint is required to calculate the surface area. For example, the roof footprint and surface area of a gabble roof and a barrel vault roof calculation are shown in Figs. 3.1a and b. Thus, the plane projection area for the inclined roof would contribute to the harvesting system.

Fig. 3.2 illustrated architects' recommended worldwide practicing sixteen roof types, i.e., Flat, Shed, Saltbox, Butterfly, Gable, Jerkinhead, Gambrel, Mansard, Hip, Pyramid hip, Dutch Gable (Half Hip), Dickey (Gullwing Polynesian), Dome, Barrel vault, Groin vault, and Domed vault. So, the roof areas might vary due to geometric patterns, but the footprint area of all roofs remains the same. The roof surface area is always more extensive than the footprint area of the rooftop.

Apart from the roof, a vertical wall can also contribute to the harvesting process, and for high-rise buildings, this would be an essential consideration. So far, among the available building code, BNBC (2020) for Bangladesh suggested that half of the attached vertical wall would contribute to rainwater accumulation.

Example problem 3.1 A two-storied building having a gable and shed roofs with a front porch. The roof footprints for the gable roof, shed roof, and front porch are 300 m^2, 51 m^2, and 9 m^2. Respectively. Thus, the relevant dimensions are AB = 20 m; CD = 4 m, CF = 15.04 m, EF = 4 m, EG = 6 m, GH = 4 m, GI = 5 m. what would be the potential catchment areas for this building?

Solution:

Thus, the available roof footprint for potential rainwater harvesting includes three types of roof composition, i.e., gable roof, shed roof, and flat roof, comprises of roof footprint and 50% of the adjoining vertical walls (Fig. 3.3).

Catchment area for downspouts #1 (DS1) = $\frac{1}{2}$ Gable roof = $\frac{1}{2} \times 300\ m^2 = 150\ m^2$.

Catchment area for downspouts #2 (DS2) = 150 m^2.

Catchment area for downspouts #3 (DS3) = Area CDFE + $\frac{1}{2}$ Area AKED.

$= (4 \times 15) + \frac{1}{2} \times (15 \times 3) = 82.5\ m^2$.

Catchment area for downspouts #4 (DS4) = Area GHJI + $\frac{1}{2}$ Area EFHG

$$= (4 \times 5) + \frac{1}{2}(4 \times 6) = 32\ \text{m}^2$$

Therefore, the potential catchment areas for this building $= 150 + 150 + 82.5 + 32 = 414.5\ \text{m}^2$.

The obtained 'rainfall yields' or 'runoff' depend mainly on catchment material and its geometry. The thumb rule states that *one-liter runoff from each mm of rainfall generated on a catchment area of 1 m² site*. Loss models for permeable or green roofs

Table 3.1 Recommendation for acquiring catchment areas as per codes around the world

Country	Code	Recommendation size or equations of catchment area
Australia	(i) National Construction Code (ABCB 2019b) (ii) Plumbing and drainage (AS/NZS 2018)	• Buildings with more than 50 m^2 should adopt rooftop rainwater harvesting
Bangladesh	Bangladesh National Building Code (BNBC 2020)	• Buildings with more than 300 m^2 should adopt rooftop rainwater harvesting; • The sloping roof catchment area includes surface area and 50% of the adjoining vertical walls
Canada	(i) National Building code of Canada (NRC 2015) (ii) Design and installation (CSA 2006) (iii) NSF Protocol P151 (NSF 2016)	• The Canadian authority maintains a tabular relationship between catchment area and storage volume
India	IS 15797:2008 (Indian Standard 2008)	• State-wise they are practicing law enforcement for different roof sizes; • The required catchment area is the tank's volume by dividing the average rainfall volume per unit area during wet months and multiplying this with the runoff coefficient • Indian standards also follow tabular relations between roof catchment area and storage water availability
UK	BS EN 16,941-1(BSI 2018)	• A graphical relation between the catchment area (roof area) and storage capacities depends on annual rainfall followed by British standards
USA	ARCSA/ASPE 63 (IAPMO and NSF 2013)	$\text{Surface area(sq.ft)} = \text{precipitation density(inches)} \times \text{demand(gallons)}/0.623 \times \text{system efficiency}$

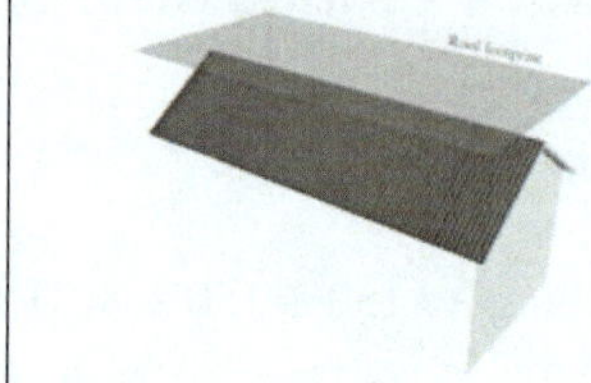	
Roof footprint, $A_{footprint} = x \times y$	Roof surface area, $A_{roof} = \frac{A_{footprint}}{cos\alpha}$
(a) Footprint and surface area of gabble roof	
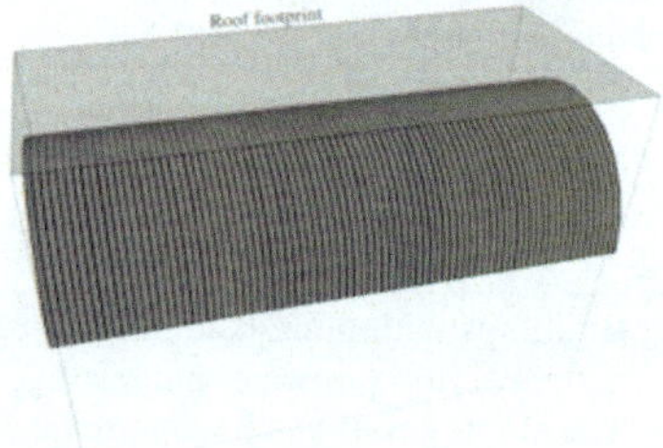	Roof footprint, $A_{footprint} = x \times y$
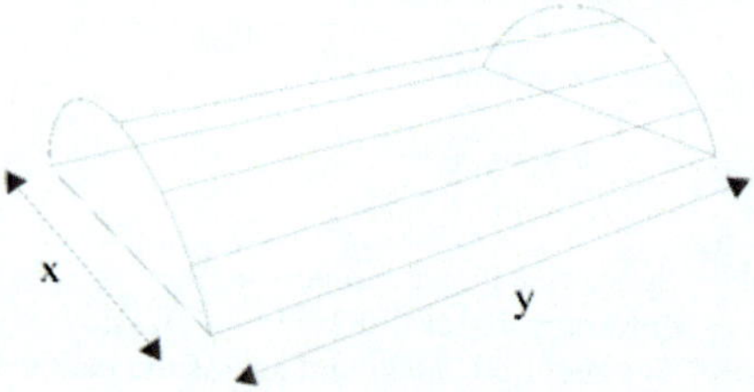	Roof surface area, $A_{roof} = \frac{\pi x}{2} y$
(b) Footprint and surface area of barrel vault roof	

Fig. 3.1 Footprint and surface area of gabble and barrel vault roof

might fail to offer this runoff amount. There could be rain shadow due to vegetation, neighboring buildings, self-orientation, winds, etc. so, a runoff coefficient is engaged while estimating the desired yield. *Runoff coefficient* is the ratio between the rainwater yield on the catchment area from a storm event and the actual rainwater delivered via conveyance. Table 3.2 presents the runoff coefficient for different types of roofs. Green roofs might have a lower runoff coefficient in a water smart city due to their hydraulic runoff characteristics.

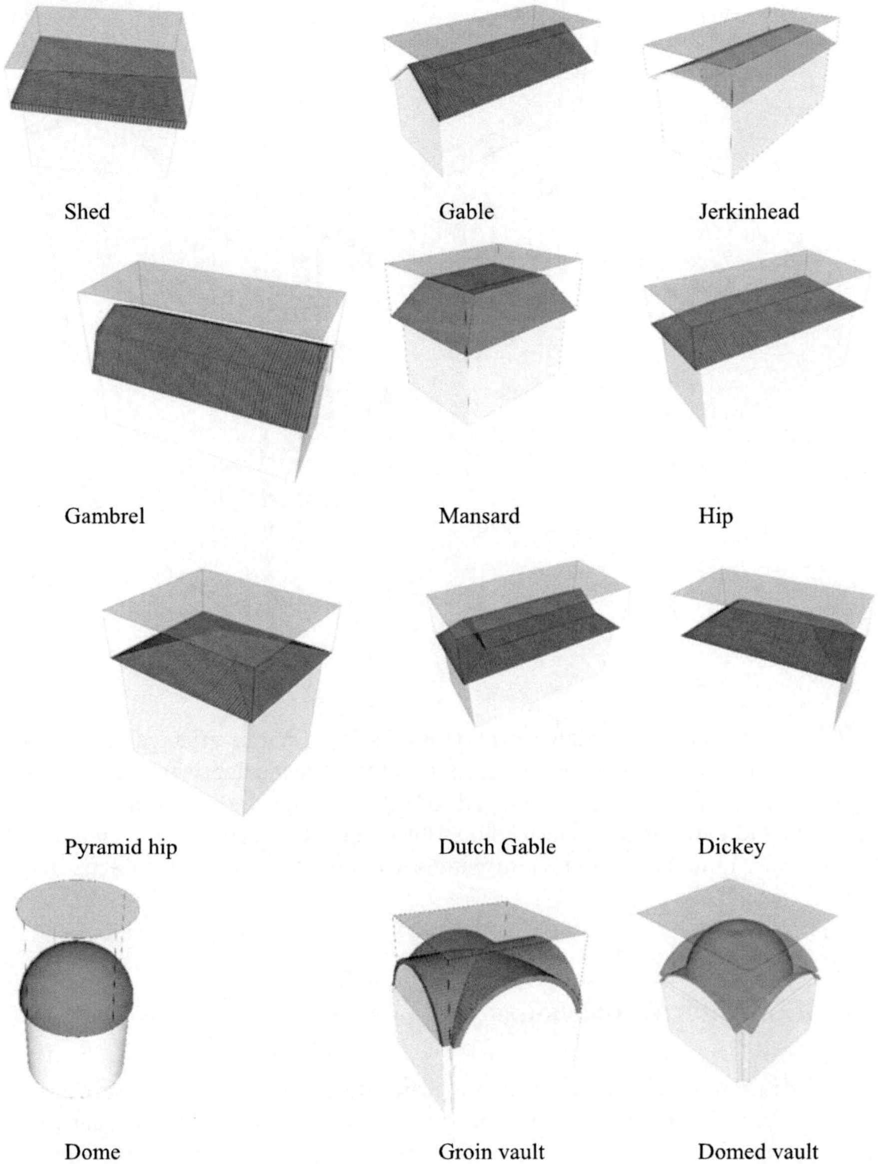

Fig. 3.2 Footprint of different roof types

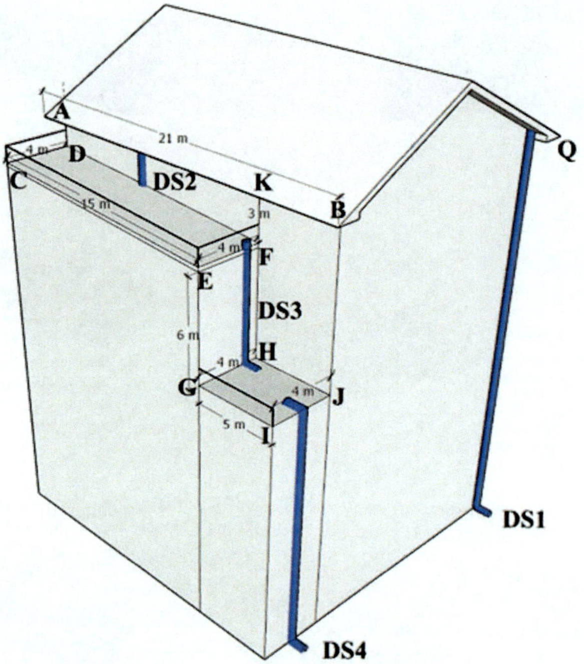

Fig. 3.3 Determining catchment area

Rainwater quality in a rainwater harvesting system can be affected in two ways: directly and indirectly. Direct contaminants include dirt and debris from atmospheric deposition, overhanging plants, bird and rodent droppings. Indirectly, the roof material could add particulate matter and dissolved chemicals. Treatment is required to ensure desired quality for the end-users discussed in greater detail in Sect. 3.7 (in this chapter).

3.3 Rainwater Conveyance

Collected rainwater from the catchment area transfers to the storage or cistern through the 'conveyance network'. This system comprises roof drains, overflow drains, scuppers, gutters, and downspouts (Fig. 3.4). These meet the quality and quantity, leaf diverters, downpipe diverter, expansion and gutter outlets, and first flush diverter. Depending on catchment geometry and materials, the conveyance system builds either penetrate the roof or are exposed as exterior edges. Transferring rainfall usually from sloping roofs is done by outer edge setup. For flat roofs, parapet walls should prevent rainwater free-fall and an intermediate drain needs for the conveyance system. The conveyance transfers rainwater from the catchment area to the storage tank by gravity or siphons action. The passage avoids stagnation as well as contamination by other sources. Considering layout physical properties, size, aesthetics, and the tank

Table 3.2 Runoff coefficients for different roof types

Type	Yield coefficients	Source
Roofs (in general)	0.7–0.9	Pacey and Cullis (1989)
	0.75–0.95	ASCE and WFCP (1969)
	0.85	McCuen (2016)
	0.8–0.9	Fewkes (2000)
	0.8	Ghisi et al. (2009)
	0.8–0.95	Lancaster (2006)
Sloping roofs		
• Concrete/asphalt	0.90	Lancaster (2006)
• Tiles	0.8	BS EN 16,941-1(BSI 2018)
• Metal	0.81–0.84	Liaw and Tsai (2004)
	0.90	BS EN 16,941-1(BSI 2018)
	0.95	Lancaster (2006)
• Aluminum	0.7	Ward et al. (2010)
Flat roof		
• Without gravel	0.8	BS EN 16,941-1(BSI 2018)
• With gravel	0.8	
Green roof [a]		
• Intensive	0.5	
• Extensive	0.7	
Permeable pavement [a]		
• Granular media	0.7	
• Plastic crates	0.8	

[a] due to uncertainties in permeable roofs, UK standard, suggested 20% standard deviations while estimating runoff yields

setup, there are two conveyance systems, i.e.: (i) wet system and (ii) dry system. The wet system comprises a pipe network to hold water after a storm, thus stopping the water drain to the tank. The pipes must be fitted with screens at each entrance and breeding insects within the pipe network.

On the contrary, the pipe network in a dry system drains out and turns dry after the storm. With larger catchment areas practicing a dry technique is difficult. However, slightly sloping catchments using a 'first flush' water diverter can offer a dry system.

3.3.1 Gutters

Channel placed around the edge of a sloping roof to collect water from the catchment and drain to the storage tank. Gutters sizing is based on the expected runoff design.

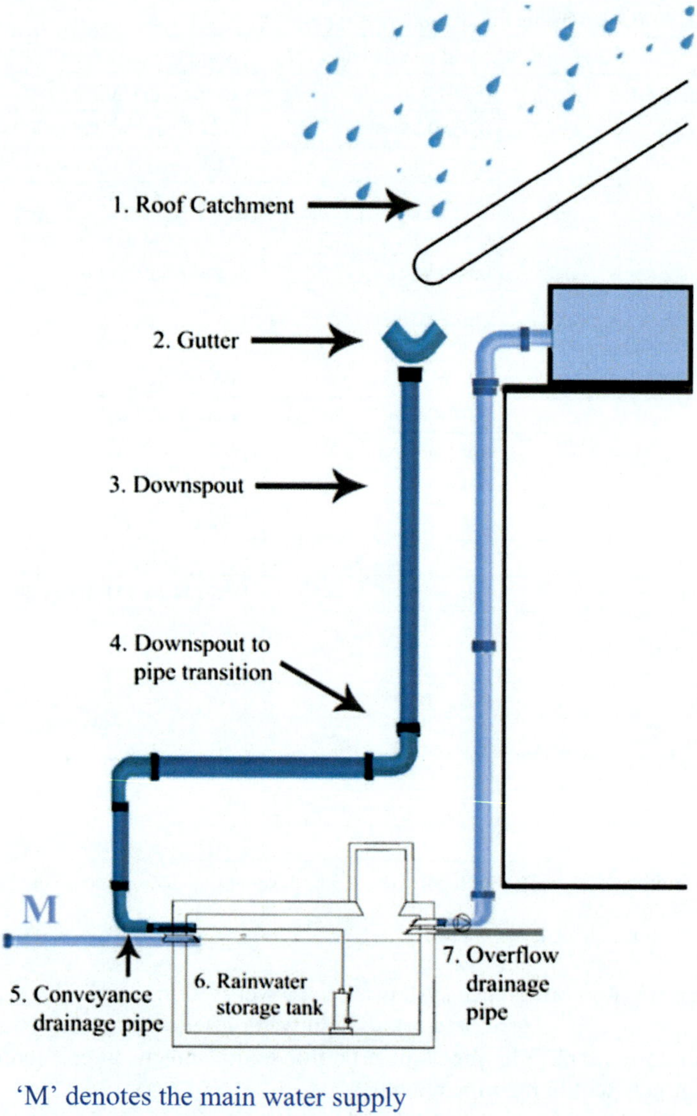

Fig. 3.4 Conveyance network for an underground rainwater storage tank

Gutters can be semi-circular, 'V' shaped, 'U' shaped, or rectangular and generally made of aluminum or galvanized steel and locally available materials plain galvanized iron sheets, polyvinyl chloride (PVC) material, etc. While quality concerns, material selection for gutter should avoid hazardous metal, wood, and plastic. Following design considerations for installing gutters are recommended based on available codes (Worm and Hattum 2006; Despins 2012):

- The design rainfall intensity for a 10-year return period and 5-min duration;
- Gutters inclined in the direction of rainwater storage tank spend shorter transfer time;
- A minimum gutter slope of 0.5–3% is recommended throughout the gutter length; and
- Gutter width includes a distance between eaves and roof edge. Gutter size for the roof drainage area calculated as:

$$\text{Roof drainage area}\left(\text{m}^2\right) = \textit{Gutter length}\,(\text{m}) \times \textit{Width}\,(\text{m}) \tag{3.1}$$

- A uniform cross-section recommends throughout the gutter length. The gutter should be 10 to 15% oversize to ensure free flow.
- Gutters supported by hangers spaced at a maximum of 450 mm.

3.3.2 Downspouts

Downspouts, also known as *downpipes* or *leaders*, refer to the vertical or inclined pipes that collect rainwater from the gutters and then convey it to storage with the provision of roof washers. The downspouts position considering the aesthetic view, physical properties, and appearance of the building. Usually, materials used for downspouts are the same as gutters and with minor cross-sections than the gutters. Installation consideration should follow as (Indian Standard 2008; Despins 2012):

- At least 100 mm diameter circular-type downspouts or 50 mm × 75 mm rectangular-type downspouts or 75 mm × 75 mm square-type downspouts are recommended;
- The number of downspouts is defined as:

$$= \frac{\text{Catchment area}}{\text{Max catchment area served by a single downspout}} \tag{3.2}$$

- A few countrywide specific charts on maximum roof drainage area served per downspout for the storm's different return periods.
- The single downspout should convey collected water up to 15 m gutter length.
- Maximum downspout offsets should be limited to 3 m.
- Minimum two downspouts need for any independent roof surface.

For planning, designing, and installing a conveyance system, the following issues should be taken into consideration:

- Selection of materials: for each network component, materials should be selected to ensure suitability for ultraviolet (UV) light exposure, burial, and rainwater quality as per the regional codes and regulations. In this regard, the manufacturer uses aluminum, galvanized steel, or polyvinyl chloride (PVC).

- Size, slope, and placement: to promote rapid water transfer towards storage, all components of the conveyance network are supposed to be sized appropriately and sloped following the catchment area. Size selections of the pipes and associated parts of the network based on the roof allocations for the collection system. Thus, the catchment area needs to be in several sections, and there should have at least two drainage areas regardless of the types of roofs. Therefore, multiple drainage pipes connected to a more extensive pipeline would transfer collected rainwater.
- Site layout and storage tank position: some catchment sections might face difficulties connecting with the conveyance system due to the grading or site layout. Architectural design should consider this issue while planning for a complex roof and remote storage. For underground storage, conveyance network design should consider burial depth and pipe slope. Similarly, an inspection should be performed to ensure other underground utility service lines (gas, water, electricity, phone, etc.) and assess the performances of planned buried pipelines for conveyance.
- Extreme weather: the conveyance network should handle the rainwater even in severe storms with proper water drainage. During freezing weather in winter, the temperature often drops below freezing (0 °C), and rainwater could be freeze with the outdoor conveyance network.
- Rainwater quality: leaf and debris diversion must be in the conveyance system to facilitate water quality in the storage tank. Treatment requires before entering storage, while the catchment area comprises a roof garden or overhanging foliage. Access to birds or insects needs to be restricted, and the conveyance network must be structurally sound-free from leakage.

3.4 Rainwater Storage

'Rainwater storage', also known as 'Rainwater cistern' or 'holding tank', is a reservoir used to store harvested rainwater collected through catchment areas using a conveyance network system. A rainwater storage system includes the following components (Fig. 3.5):

(1) Conveyance drainage pipe as an inlet for rainwater storage:

Through this inlet, collected water, after transferring through the conveyance system, enters the tank.

(2) Tank:

Tank is the reservoir for harvesting rainwater: tank capacity, placement, and material influence quantity and quality. Multiple tanks, linked at the top or the bottom, can be connected to increase storage capacity.

(3) Smoothing inlet:

It is also known as a 'calming inlet', usually made of stainless steel and used to remove turbulence of the incoming water.

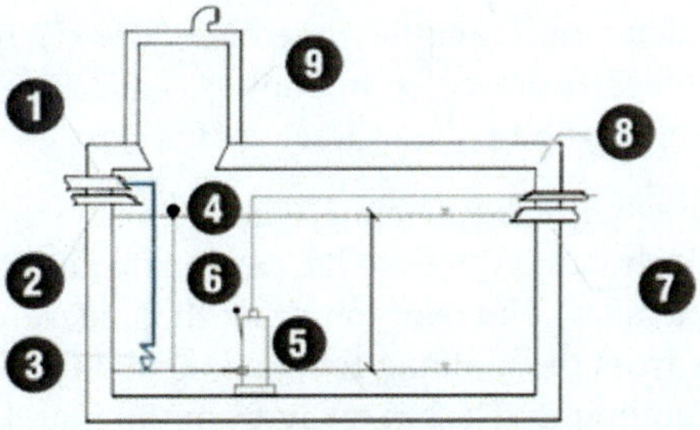

1: Conveyance drainage pipe as an inlet for rainwater storage; 2: Top-up drainage pipe; 3: Smoothing inlet- stainless steel 'flow calming'; 4: Floating stainless steel suction filter; 5: Submersible feed pump; 6: Low water cutoff float switch; 7: Overflow pipe; 8: Storage Tank; 9: Access riser (underground storage) or access hatch (overground storage)

Fig. 3.5 Components of rainwater storage

(4) Water level indicator:

This device is used to monitor water levels within the tank and communicate with distribution components. Monitoring might use floats or electronic-based sensors.

(5) Pump or pump intake:

A submersible feed pump within the rainwater storage tank capacity requires for the extraction of stored water. Beyond this capacity, i.e., more than the higher water level or lower than the low water level, harvested water might not meet the design water quality while extracting. A floating screen inlet reduces the vortex and introduces air into the pumping system.

(6) Overflow drainage pipe:

Excess water passed away from the tank through the drainage pipe.

(7) Air vent:

Provision for an air vent is required to release air while the tank is filling. Alternatively, the overflow pipe can serve as the vent.

(8) Tank access

For maintenance, underground tanks need an 'access riser' and 'access hatch' is required for the above-ground and intermediate tanks (Fig. 3.5).

3.4.1 Tank Materials and Location

Based on availability and regional practices, concrete, plastic, fiberglass, etc., are used to construct rainwater storage tanks. On the other hand, site topography, size, and shape tanks can be located at the ground or underground or directly integrated into a building. Thus, storage can be found as overhead storage, rooftop storage,

storage at an intermediate level, and the basement floor (Haq 2016). A location for the selected storage place is presented in Table 3.3 and Fig. 3.6. Thus, the type of storage selection is influenced by the following factors:

- Purpose of the storage: Potable or non-potable
- Feasibility: storage should be close to the catchment and main distribution line to reduce cost and leakages. The overflow from the underground and ground-level storage tanks pass from the building foundation, and the available recommendation suggested a minimum of 1.2 m away from the foundation (Indian Standard 2008). Underground storage in complex rock areas is not recommended.
- Location: storage tanks should be located in shaded places to reduce contamination considering building layout; however, avoid dense tree cover.
- Weather and climate: material selection made considering the adjacent environment features. In cold countries, storage in-ground, as well as above-ground, might be at risk for freezing. Galvanized Iron material might fail to offer the desire performance in coastal areas.
- Budget allocation, material availability, and design tank size influence the storage selection.

3.4.2 *Tank Sizing*

The designed storage tank capacity needs to be as per the water demand and water availability. The water demand includes the number of consumers, per capita requirement, and dry days requirement. On the other hand, water availability is computed based on catchment type, average annual rainfall, number of dry days, and rainfall pattern. Various computation approaches on tank sizing are available worldwide (Table 3.4). Overall, three primary considerations on tank sizing are:

- *A simplified approach or demand-side approach* is to provide a specific volume of water during dry periods with available supply;
- *The intermediate approach or supply-side approach* is to provide a specific volume of water with limited supply (mass curve technique); and
- *A detailed approach* ensures optimum tank size to a minimum water budget able to maximize water savings.

Simplified approach or demand-side approach: Tank size is determined by the per capita demand, number of water users, and desirable harvested water use time. This approach recommends in arid climates with a distinct dry season. If rainfall is more than the demand, then size the tank as per water demand. Typically, along with this calculated volume, an extra water volume is provided as a freeboard, which varies between 10 and 25% (CMHC 2012; BNBC 2020). As the assumptions relating to consistent daily demand are absent, not recommended for commercial buildings.

Example problem 3.2 Due to groundwater lowering, the city area implemented outdoor watering bans for summer. Among the households, a household of four

Table 3.3 Comparison among rainwater storage in different locations

	Underground	Ground level (away from the building)	Overhead	Rooftop	At intermediate level	In basement floor
Advantages	• The absence of daylight reduces algal growth • They are protected from weather conditions, i.e., freezing or warming • Space saves on the building premises	• Low cost for ground excavation • No groundwater-related issues • Usually less expensive to install • Less chance of flood inundation • Easily accessible for repair or inspection	• Adequate pressure within the tank intake and outlet • Easily accessible for repair or inspection	• The intake pipe length for collection is supposed to be shorter • Water transfers to the rooftop tank by gravity • Collection pipe might not be required	• Collected rainwater from the catchments conveyed to the tank under gravity • Easy access could ensure frequent water quality monitoring • Ensure desired pressure in the supply system	• The absence of daylight reduces algal growth • Construction as an integral part might improve structural integrity
Disadvantages	• Lesser accessibility for monitoring • Higher installation cost due to excavation requirements • An empty tank is prone to uplifting risk • Location needed to be free of buried service lines	• Risks involved in different weather events, i.e., freezing or heating • Additional space occupies within the premises • It might be aesthetically undesirable	• Exposure to sunlight • Risks involve freezing or heating • Conflicts with other overhead utilities and future vertical extension • Obstructs building aesthetical view • Structural safety ensures as an integral part of building • Height restriction persists for the building	• An extra load needs for the structural design of the building • Earthquake severity and height restriction might disrupt regional building codes • Structural design should manage slosh-dynamic problems	• Reduces functional floor space of the building • An additional load needs for the structural design of the building • Noise pollution due to pumping in the adjacent floors • Maintenance activities might be delayed compared to other routine activities	• Extra pumping and overflow drainage facilities are required • Leakage from the reservoir would weaken the load-bearing properties of the building foundation soil

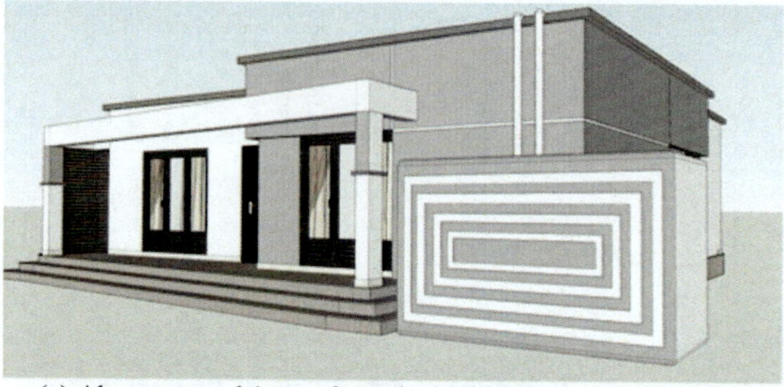

(a) Above ground (away from the building) rainwater storage

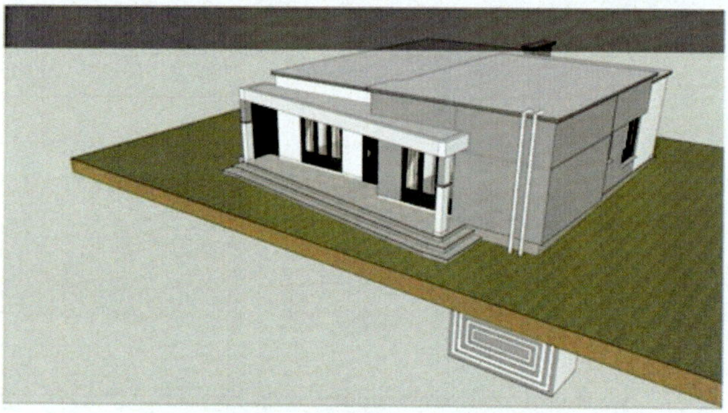

(b) Underground rainwater storage

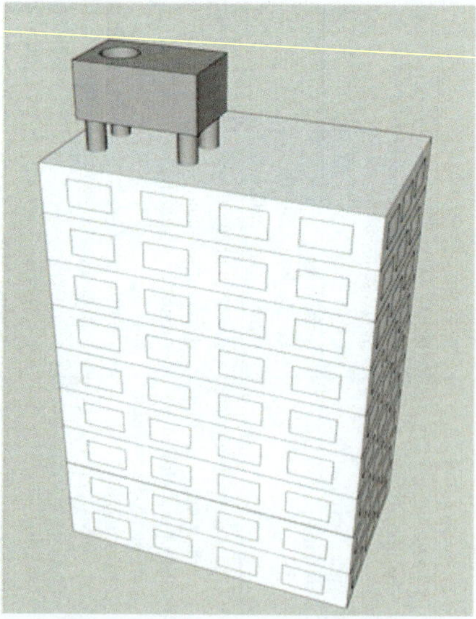

(c) Overhead rainwater storage

Fig. 3.6 Different locations of rainwater storage

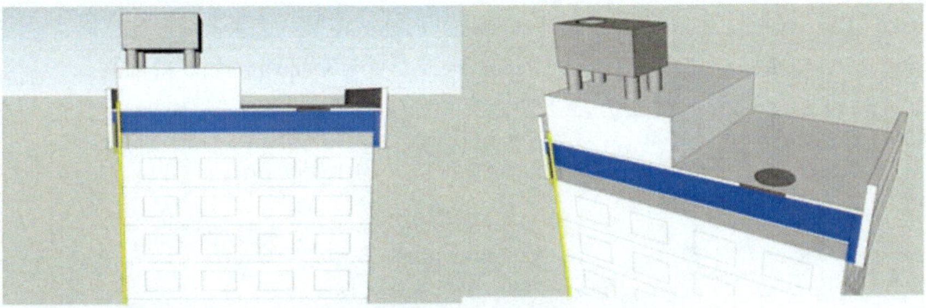

(d) Rooftop rainwater storage

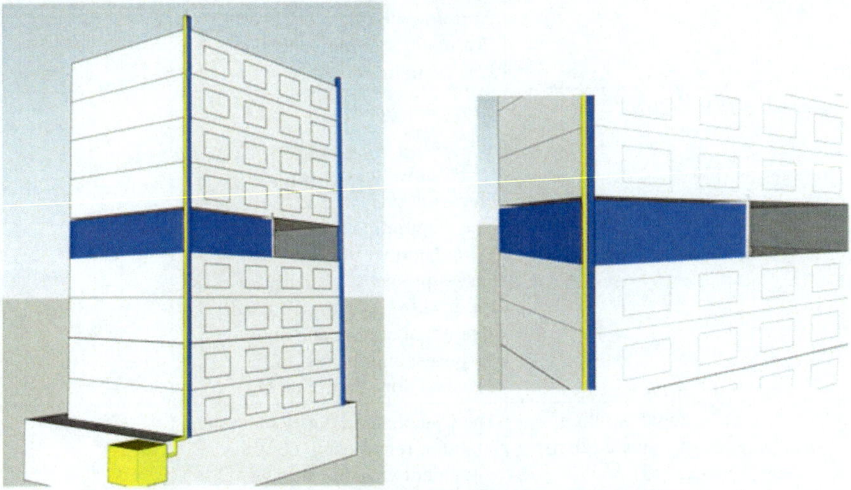

(e) Rainwater storage at an intermediate level

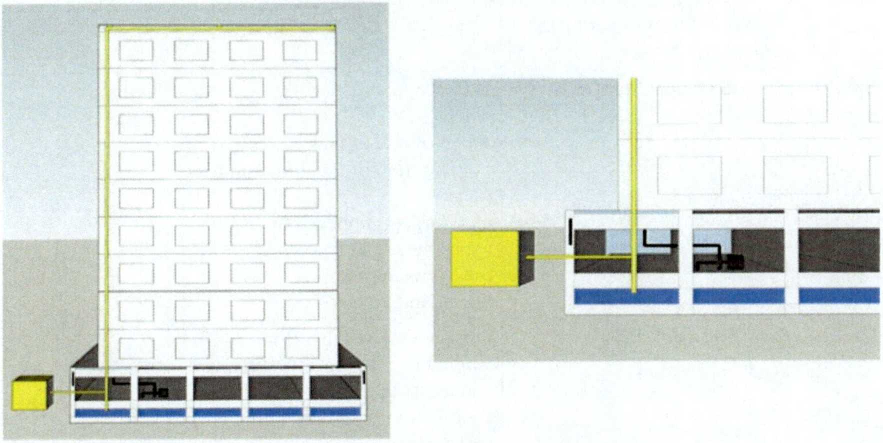

(f) Rainwater storage in the lowest basement floor

Fig. 3.6 (continued)

Table 3.4 The recommended size of the storage tank by different countries

Country	Code	Recommended size or equation of storage capacity	Remarks
Australia	(i)ABCB (2019b) (ii)AS/NZS 3500 (AS/NZS2018)	*Runoff* (L) = $A \times (rainfall - B) \times roof\ area$ Here, A = collection efficiency, a typical value is within 0.80–0.85 (Nath et al. 2006) B = associated loss due to absorption, and the typical value is 2 mm/month (Nath et al. 2006) 'Rainfall' is in mm, and 'roof area' is in m^2	–
Bangladesh	BNBC (2020)	$Storage\ volume(m^3) = \frac{D \times N \times Dp}{1000} + Floating$ D = Rainwater demand, liter/capita/day N = Population number D_p = Number of rainwater harvesting days. Usually, 90 days for drinking, cooking, utensils, cleaning, bathing, and ablution purposes; and the rest of the 210 days for other purposes	–
Canada	(i) National building code of Canada (2010) (NRC 2015) (ii) CSA standard B128.1(CSA 2006) (iii) NSF protocol P151 (NSF 2016)	The Canadian authority maintains a tabular relationship between catchment area and storage volume	–
India	IS (Indian Standard 2008)	$V = t \times n \times q$ where V = Volume of tank, litres; t = Duration of the dry period (days); n = Target number of the consumers; and q = Consumption, litres/capita/day	Dry season demand
UK	BS (BSI 2018)	Storage tank volume (litres) = Annual rainfall (mm) × Effective collection area (m^2) × Drainage coefficient (%) × Filter efficiency (%) × 0.05	

(continued)

Table 3.4 (continued)

Country	Code	Recommended size or equation of storage capacity	Remarks
USA	ARCSA/ASPE 63 (IAPMO and NSF 2013)	*Runoff* (*Gallons*) = $A \times (Rainfall - B) \times Roof\ Area$ A = Collection efficiency, the typical value is within 0.80–0.85 (Nath et al. 2006) B = Associated loss due to absorption; typical value is 0.08 inches/month (Nath et al. 2006) *Rainfall* expressed in inches and *Roof Area* in sq. feet Then, $V_t = V_{t-1} + (\text{Runoff} - \text{Demand})$ where V_t = Theoretical volume of water remaining in the tank at the end of the month V_{t-1} = Volume of water left in the tank from the previous month	Similar to Australian code

planned a rainwater harvesting system to cover the outdoor watering needs of potted plants. The estimated per capita outdoor watering is 27.9 liter/day, and the available rainfall data confirms to meet these requirements. In this city, oversize factors of 1.25, i.e., 25% larger than the calculated tank size, are usually applied to ensure that the tank is not full of water when dry periods start. Determine the required volume of the storage tank.

Solution:

Assuming a 100-day dry period.

Estimated water consumption $= 27.9 \times 100 \times 4 = 11160$ liter.

Oversized factor = 1.25.

Recommended tank size $= 11160 \times 1.25 = 13950$ liter(round up to 14, 000 liter).

A storage tank of 14 m^3 should provide enough rainwater to last for outdoor watering during the summer for a household of four.

Intermediate approach or supply-side approach: If rainfall is less than the demand, size the tank to rain availability and plan for the scarce or the critical period. This approach applies the mass curve technique with a dataset of monthly harvested rainwater, monthly demand, and the difference between these two is the monthly storage. The storage volume for a particular year estimates from the difference between the minimum amount stored during the dry season and the maximum amount stored during the wet season. Thus, the difference between the maximum and minimum amounts stored each year and the most significant difference yields the tank size (Example problem 3.3). This approach is apparent that the variation of rainfall would affect the overall system efficiency.

Detailed approach: This approach is applicable for variable demands and uncertain yields throughout the year and is also planned for a large storage tank. Also, practices in the stormwater management while integrated into the rainwater harvesting system. In this approach, there are three methods to determine the storage capacity (BSI 2018):

- Probability analysis with at least 5-year time series:

  Real rainfall dataset involves providing information in the range of various return periods. This method aims to balance the extra storage offered by the rainwater harvesting system and the existing stormwater drainage system.
- Analysis of at least 20 extreme events:

  The observed rainfall dataset should include hourly information, and each event supposes to offer at least three months of antecedent rainfall. The return period of each event should be computed in the site location context.
- Based on the 100-year extreme stochastic series:

  Analysis of the 100-year series would cover all significant events, including numbers of events, dry periods, intensities, etc.

Due to global warming, considering the temporal and spatial fluctuations of rainfall data as a part of climate changes, there are advancements in demand and supply-based approaches. Rainwater storage design tools and tables determine site-specific optimal storage tank capacity. Design tools are also available for a wide range of usage and researches going on. For instance, based on queuing theory, the Moran model (Nagy et al. 2002) develops to determine optimum tank capacities of rainwater harvesting. In this model, simultaneous computations relate to reservoir capacity, demand, and supply. This model experiences difficulties with handling critical period and affect the overall system efficiency. The *behavioural model* has been using by researchers (Schiller and Latham 1987; Fewkes 2000).

Example problem 3.3 A seven-storied building, the ground floor dedicated to parking, and 32 occupants reside in 6 apartments (Fig. 3.7). The daily per capita water demand is 120 liters. The available roof footprint is 234.1 m^2. The mean annual rainfall is recorded as 2918.1 mm. The wind is most often from the south direction for around six months, starting in March. The owner of this building is planning to adopt rooftop rainwater harvesting. The historical rainfall distribution is found as:

Month	Jan	Feb	Mar	Apr	May	Jun	Jul	Aug	Sep	Oct	Nov	Dec
Rainfall (mm)	5.6	24.4	54.7	147.4	298.6	607.3	727	530	259.3	184.4	67.5	11.9

(a) Front view of the building

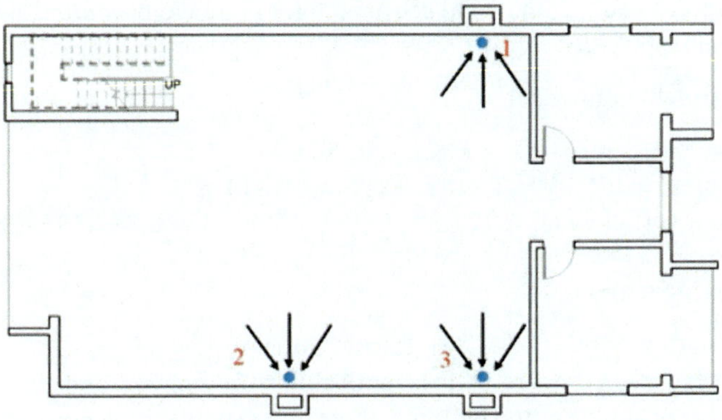

(b) Rainwater downspouts on roof

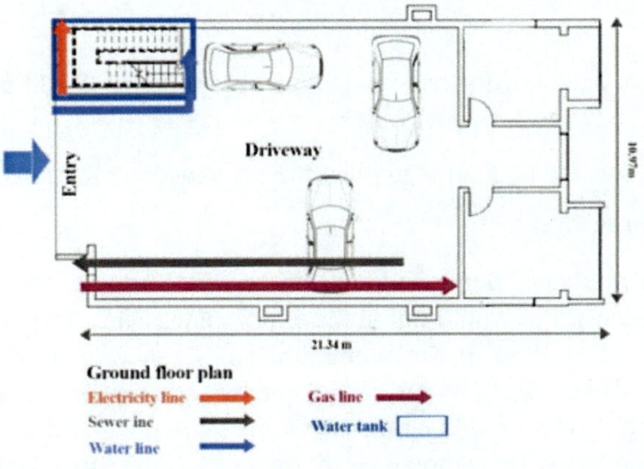

(c) Ground floor plan

Fig. 3.7 Residential building for Example 3.3

What rainwater storage size would you recommend for this building?

Building details:
Building height = 31.1 m.
Building footprint area = 234 0.1 m^2.
Open area: Paved area = 105.91 m^2; Unpaved area = 20.90 m^2.
Contribution of the open areas as well as verticle walls is absent.

Solution:

Step 1: Demand estimation.

Total demand: $32 \times 120 = 3840$ L per *day* = 128 m^3 *per mean month.*

Step 2: Supply estimation.

Using Tables 2.3 and 3.2, the yield coefficient for concrete pavement and flat roof as 0.7 and 0.8, respectively.

Annually available rainwater

a. Roof $= 234.1 \times 2.9181 \times 0.8 = 546.50\,\text{m}^3$
b. Paved area $= 105.91 \times 2.9181 \times 0.7 = 216.30\,\text{m}^3$
c. 50% of a verticle wall $= \left(21.34 \times \frac{31.1}{2}\right) \times 2.9181 \times 0.7 = 677.83\,\text{m}^3$

Daily available water

Option I: Rooftop only $= \frac{(546.50)}{365} = 1.5\ \text{m}^3/day$ i.e., 45 $\frac{\text{m}^3}{mean\ month}$

For conveyance, compare to the applicability of Eq. 3.2 and the architectural features and appearance of the building, three downspouts provisions selected. Here, the downspouts are 50 mm × 75 mm rectangular-type shown in Fig. 3.7b.

Option II: All possible sources $= \frac{(546.50+216.30+677.83)}{365} = 3.95\ \text{m}^3/day$ i.e., 118.5 $\frac{\text{m}^3}{mean\ month}$

In addition to the three downspouts (Fig. 3.7b), gutters and additional downspouts provisions are required for conveyance.

Step 3: Sizing storage tank aims to meet the dry season demand versus supply.

Option I: Rooftop only

If the planned rainwater harvesting system aimed to supply water throughout the year, the available water would fail to the expected demand. The tank needs to meet the water demand of 45 m^3 per mean month. The water year starts in April; assume that the tank will be empty by the end of the dry season, i.e., March. Since April, the harvested rainwater could begin to meet the demand, and the maximum surplus occurs in September with a storage requirement of 208.23 m^3 (Table 3.5).

So, the storage tank size $= 208.23 \times 1.1 = 229.053$ (round up to 230 m^3). The ground floor is using as parking; then an underground storage tank needs to be built. The tank has been planned to place under the stairs considering the associated utility

Table 3.5 Computation of storage tank requirements

Month	Rainfall (m)	Harvested rainwater (m^3)	Cumulative harvested rainwater (m^3)	Demand (m^3)	Cumulative demand (m^3)	Deficit (m^3)
Apr	0.1474	27.61	27.61	45.5	45.5	-17.89
May	0.2986	55.92	83.53	45.5	91	-7.47
Jun	0.6073	113.74	197.26	45.5	136.5	60.76
Jul	0.727	136.15	333.41	45.5	182	151.41
Aug	0.53	99.26	432.67	45.5	227.5	205.17
Sep	0.2593	48.56	481.23	45.5	273	**208.23**
Oct	0.1844	34.53	515.77	45.5	318.5	197.27
Nov	0.0675	12.64	528.41	45.5	364	164.41
Dec	0.0119	2.23	530.64	45.5	409.5	121.14
Jan	0.0056	1.05	531.69	45.5	455	76.69
Feb	0.0244	4.57	536.26	45.5	500.5	35.76
Mar	0.0547	10.24	546.50	45.5	546.5	0.00

connections (Fig. 3.7c). In this case, a depth of 7.5 m has been suggested following the available dimensions, i.e., length and width are 9 m and 3.5 m, respectively.

Option II: *All possible sources* might supplement the demand requirements around three times more than option I. And if needed, another verticle wall includes the harvesting process—similarly, tank size and position based on the harvested water use mode. Again, to use the rainwater only during rainy days, some storage pots are needed to preserve average daily rainwater on the peak rainfall month for everyday use. Temporarily these storage placements are in the ground floor parking area or over the roof.

3.4.3 Storage Performances

Rainwater storage performances are investigated through three approaches, i.e., (i) Behavioural model, (ii) Water-saving efficiency, and (iii) Generic curves for system performance.

The **behavioural model** is based on two fundamental algorithms, i.e., the **Y**ield **A**fter **S**pillage (**YAS**) and the **Y**ield **B**efore **S**pillage (**YBS**) (Jenkins et al. 1978). These operating rules illustrate by a typical rainwater conveyance system shown in Fig. 3.8.

The **YAS** operating rule allocates the yield as the smaller amount between stored rainwater earlier and the demand at present (Eq. 3.3). Thus, the present accumulated rainwater volume is the total rainwater volume stored earlier and the excess spilling and deducts the current yields (Eq. 3.4). Thus, the YAS operating rule is:

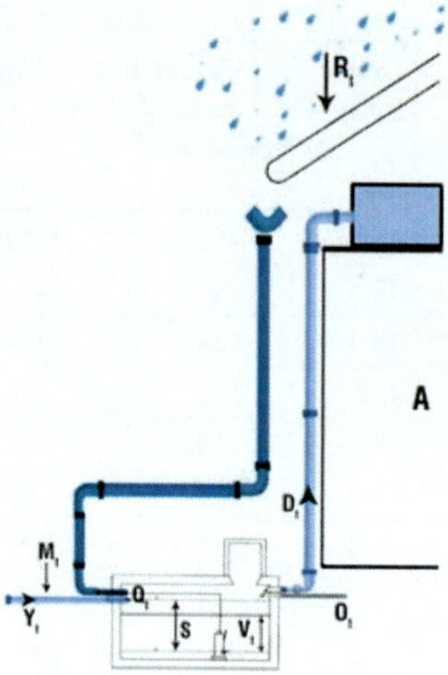

Where, during the time interval, t
Rt = Rainfall
Dt = Demand
Yt = Yield
M_t = Mains water
A = Catchment area
Vt = Volume in store
S = Storage capacity
Qt = The roof runoff
Ot = The overflow

Fig. 3.8 Typical rainwater conveyance system

$$Y_t = \min\left[D_t t;\ V_{t-1}\right] \tag{3.3}$$

$$V_t = \min\left[(V_{t-1} + Q_t - Y_t);\ (S - Y_t)\right] \tag{3.4}$$

The **YBS** operating rule allocates the yield as the smaller amount between stored rainwater in the past and present, and the current demand (Eq. 3.5). Thus, the present accumulated rainwater volume is the earlier storage before deducting the yield with the excess spilling and deducts the current yields (Eq. 3.6). Thus, the YBS operating rule is:

$$Y_t = \min\left[D_t;\ V_{t-1} + Q_t\right] \tag{3.5}$$

$$V_t = \min\left[(V_{t-1} + Q_t - Y_t);\ S\right] \tag{3.6}$$

Generally, the reservoir operating algorithm in a behavioural model as:

$$Y_t = \min\left[D_t;\ V_{t-1} + \theta Q_t\right] \tag{3.7}$$

$$V_t = \min\left[(V_{t-1} + Q_t - \theta Y_t) - (1-\theta)Y_t;\ S - (1-\theta)Y_t\right] \tag{3.8}$$

where

θ is a parameter of two values, i.e., 0 and 1, representing the algorithm as YAS and YBS, respectively.

Water-Saving Efficiency (η) is the ratio of total yield through replacing mains water and the overall demand:

$$\eta = \frac{\sum_{i=1}^{i=t} Yt_i}{\sum_{i=1}^{i=t} Dt_i} \times 100\,\% \tag{3.9}$$

Based on YAS for rainwater harvesting performance evaluation, generic curves consider catchment area, storage capacity, and demand. To prepare generic curves, two dimensionless ratios, i.e., demand fraction $\frac{D}{AR}$ and storage fraction $\frac{S}{AR}$ present in graphical mode. Here, the effect of demand patterns, catchment runoff coefficient, and variation in rainfall dataset are the influencing factors. Fewkes (2000) generic curves for water-saving efficiencies showed acceptability in many countries.

3.5 Distribution System

A distribution system engaged in disseminating harvested rainwater to permitted fixtures. This system includes pumps, pressurizing, conditioning, controlling flow to end-use, monitoring storage tank levels, backup water, bypass, and makeup water.

Pumps and pressuring distribution systems are interconnected within the building through the rainwater storage system. Harvested rainwater withdraws from the *underground storage tank,* either pump placement *inside* or *outside the tank*. The recommended immersion depth maintains for the pump inside the tanks. A frost-free, well-ventilated location for the pump outside the tank is selected, along with noise and vibration-free mountings. In both cases, a non-return valve should provide an isolating valve for maintenance. Usually, to eliminate pump dry running conditions, water level sensors are often used. Pump selection for a given rainwater harvesting system considering the following criteria:

- Pump location;
- Configuration of the pump controller;
- Operating voltage;
- Pumping rate;
- Pump head; and
- Daily pumping duration.

Among the available pumping options, i.e., centrifugal pumps and positive displacement pumps, usually, *centrifugal pumps* are based on the pressurized distribution system. A pump should generate kinetic to meet the total energies, i.e., summation of energy due to elevation, friction, pressure, and velocity. A *rotating impeller* increases water pressures within the centrifugal pump. Both submersible and jet pumps are based on the working principle of underwater mounted centrifugal pumps. *Submersible pumps* have a longer lifespan and lesser space requirements for their placement but are expensive than jet pumps. *Positive displacement pumps are* practiced based on both manually automatic operations. The storage tank locations, operation, maintenance, and budget, a submersible or jet pump consider extracting water for direct supply or passed the harvested water to any intermediate storage tanks. In the *underground-overhead tank* system (Fig. 3.9a), water is stored in underground tanks and pumped for supply through the overhead tank by gravitational force. A similar approach applies for overground, i.e., water stored in *above-ground tanks* and pumped for overhead tank collection for distribution by gravity to use (Fig. 3.9b, c). Due to simplicity in operation and maintenance and cost-effectiveness, an underground-overhead tank system seems the better approach.

On the other hand, water is stored underground (Fig. 3.9a) or overground and directly pumped to the point of use. However, a *direct-pumping* system needs a low initial cost but is expensive in operation and maintenance. However, the composition of both methods is required while considering harvested rainwater would supplement the main supply (Fig. 3.9).

Rainwater conveys to underground storage as a backup with the mains supply. Harvested rainwater is pumped to an overhead tank to feed the user points through water treatment (Fig. 3.9c). On the other hand, service water from the mains supply is pumped directly to an overhead storage tank. The harvested rainwater supplements the service water, and the storage tank is at an intermediate level (Fig. 3.9d).

Two pump controller configurations are available for a pressurized distribution system: constant speed pumps and variable speed drive (VSD) pumps or variable frequency drive (VFD) pumps.

A *constant speed pump* can be activated following a significant drop in system pressure and then pumps to replenish water storage in the pressure tank.

VSD/VFD pumps can control pump impeller speed to provide the required amount of water for the pressure system.

Multiple pump systems should arrange a standby pump if necessary. Also, water hammers, surges, and hunting should avoid using a diaphragm expansion vessel.

The operating voltage would vary among countries; the type of pump is decided as per the manufacturer's recommendation, i.e., 120 V or 240 V.

Pump flow rate is the required water quantity per unit expected from pump operation. The information is necessary to determine the flow rate is the distribution system's types, sizes, and fixtures. Then, the pump size should be selected to meet the 'maximum peak flow' requirements. For pumping through an underground-overhead tank system, the pump flow rate (Q) depends on the overhead storage capacity (volume of water, V) and the required operation time (t).

$$Q = \frac{V}{t} \tag{3.10}$$

For direct pumping, the Hunters curve practices worldwide to determine the probable flow rate. Hunter's curve developed in 1940, highlighted on fixture unit-wise estimated demand, and in 1987, the ASHRAE modified Hunter curve for different

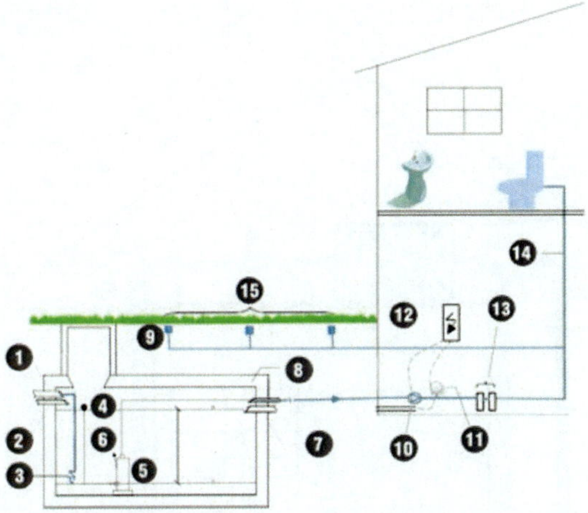

1: Conveyance drainage pipe as an inlet for rainwater storage; 2:Top-up drain; 3: Water level sensor; 4: Floating stainless steel suction filter; 5: Submersible feed pump; 6: Cut-off float switch; 7: Overflow drain 8: Storage tank 9: Riser 10: Pump 11: Pressure tank 12: Electrical supply panel 13: Post-storage treatment units 14: Harvested rainwater supply pipe 15: Sub-surface irrigation system

(a) Underground rainwater storage with direct pumping system

(b) Underground – overhead rainwater storage system

Fig. 3.9 Pumps and pressurized distribution system

(c) Rooftop storage through water treatment

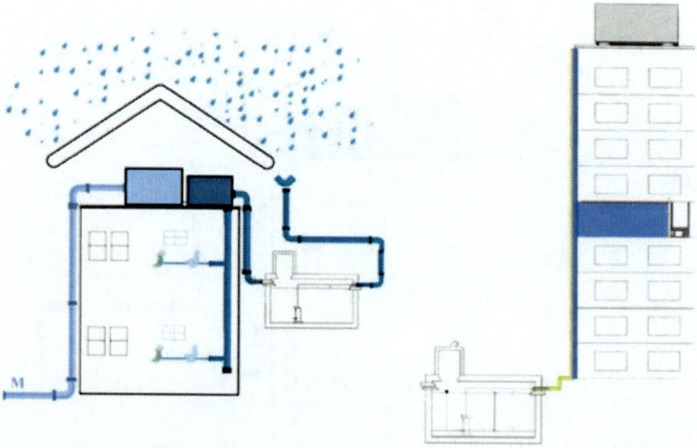

(d) Rainwater stored at an intermediate level

Fig. 3.9 (continued)

Table 3.6 Details on water supply fixture unit for household indoor fixtures (Vickers 2001)

Fixture type	Water requirement	Minimum flow rate per fixture[b] liters/minutes
Toilet • Low flush • Ultra-low flush • Dual-flush	• 13.0 liter/flush • 6 liter/flush • 4.8 liter/flush	2.7
Laundry • Top-loading • Front-loading	• 150 liter/load • 100 liter/load	19
Lavatory • Inefficient/old • Standard • High-efficiency	• 8.0 liter/minutes • 5.3 liter/minutes • 3.2 liter/minutes	1
Shower or bathtub • Inefficient/old • Standard • High-efficiency	• 9.5 liter/minutes • 8.3 liter/minutes • 5.7 liter/minutes	19
Kitchen sink	–	1.6
Diswasher	–	7.6

[b] Total usage column values ÷ 7

end-users was developed considering technological development. However, country-wise technological developments are observed worldwide for indoor and outdoor fixtures.

There are two distinct pipe sections, i.e., rainwater service pipe and rainwater supply pipe. Rainwater service pipes covey water from the storage tank to the pumps directly or through a control unit for jet and submersible pumps, respectively. Then, rainwater supply pipes convey water to the permitted fixtures (Table 3.6).

The amount of pressure or "head" is the kinetic energy a pump needs to transfer water in pumping water. To calculate the pump head, two factors, i.e., required system pressure by the connected fixtures and the total head. The total head for a moving pump or the total dynamic head (Fig. 3.10) applies to lifting underground reservoir water to an overhead tank.

$$\textit{Pump head} = \textit{Required system pressure} + \textit{Total dynamic head} \tag{3.11}$$

where

Total dynamic head = Static lift + Static height + Friction loss

Here,

System pressure is rainwater fixtures operating pressure. For typical residential applications, this might be 275–415 kPa, i.e., 27.5–41.5 m (CMHC 2012) unless there is another recommendation in the concerned rainwater harvesting code.

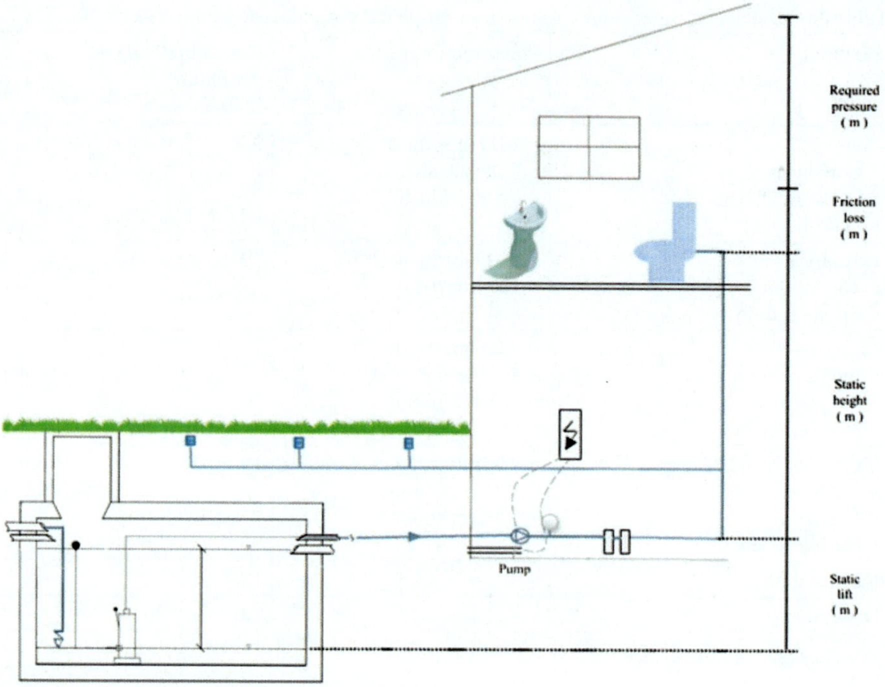

Fig. 3.10 The required pump head

Static lift = height of water to be to the pump-datum level

Static height = height of the water from the pump-datum level to the required higher level

Friction loss = pressure losses during water travels through pipes and fittings can be calculated using Hazen-Williams Equation:

$$V = 0.85C\left(\frac{d}{4}\right)^{0.63}\left(\frac{h_f}{L}\right)^{0.54} \tag{3.12}$$

where

C = Hazen-Williams coefficient.

d = Pipe internal diameter (m).

h_f = Head loss (m).

The pump has been selected based on the regionally available 'pump curves' with a known flow rate and the pump head. Pumping duration depends on the mode of water services, i.e., potable or non-potable, and the supply requirements. For instance, in an individual residential with non-potable usage, an interruption can be acceptable compared to a commercial or multi-residential area with potable use. A pressure tank can offer constant pressure within the distribution system. Sizing the pressure tank should be depending on the variation of pump size and flow rate. The required capacity for the pressure tank is as:

$$Tank\ size\ (liter) = \frac{Pump\ flow\ rate\left(\frac{liter}{\text{min}}\right) \times Pump\ run\ time(\text{min})}{Drawdown\ factor} \quad (3.13)$$

Example problem 3.4 For the seven-storied building stated in Example problem 3.2, the planned rainwater harvesting aims to serve only the non-potable purposes, i.e., three toilets and a washing machine for each floor (Fig. 3.11) and a 12 mm hose bib provision the ground floor. There is an overhead tank of 5.8 m height on top of the building height of 31.1 m. The installed rainwater service pipe and rainwater supply pipe diameter are 32 mm and 18 mm, respectively. Determine the maximum peak demand for the pump and pressure system.

Solution:

For the indoor, application of rainwater for water closet and laundry:

- The ultra-low flush toilets in the home require 6.0 liter/flush. Assume, per person requires five flushes in a day.
- The front-loading type washing machine requires 100 liter/load.

Rainwater to meet 32 persons' toilet requirements = $6 \times 5 \times 32$ = 960 liters/day = 6720 liters/week.

The weekly rainwater usage for laundry = 3 loads per family $\times$ 6 $\times$ 100 liters per load = 1800 liters/week.

Converting the weekly to the daily indoor rainwater demand (total) = $\frac{6720+1800}{7}$ = 1217.14 liters/day.

To meet the above demand by rainwater harvesting, peak demand:

Indoor fixtures	Fixtures number	Minimum flow rate (per fixture) (lpm)	Total flow rate (lpm)
Toilet	18	2.7	48.6
Washing machine	6	19	114
Hose watering (12 mm supply)		11	11
Maximum peak demand			173.6

Thus, the required minimum pump flow rate is **175 liters per minutes** $(\approx 0.0029\,\text{m}^3/\text{s})$.

As per Eq. 3.13,

$$Total\ dynamic\ head = Static\ lift + Static\ height + friction\ loss$$

The required static lift is 7.5 m (using the same storage depth described in Example problem 3.3). Then, the static height is 31.1 m (building height).

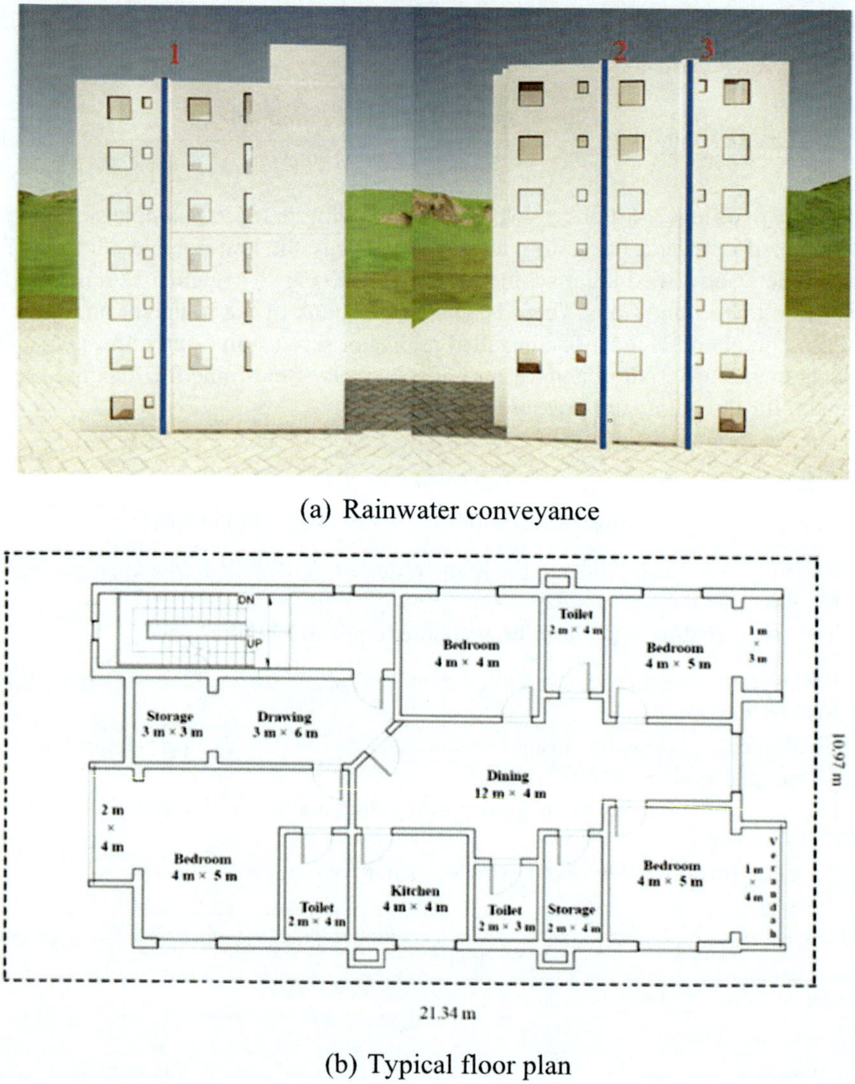

Fig. 3.11 Example problem 3.3

Total friction loss: assuming Polyvinyl Chloride (PVC) pipe as the service pipe (diameter = 32 mm) and supply pipe (diameter = 18 mm). The Hazen-Williams Coefficient from the Table 3.7 is 150. If there is only one pump:

$$Friction\ loss,\ h_f = \left[10.65\left(\frac{0.0029}{150}\right)^{1.85}\frac{38.6}{(0.032)^{4.87}}\right]$$
$$= 14.91\ \text{m}$$

Table 3.7 Hazen-Williams coefficients for pipe roughness (Williams et al. 1920)

Material	C factor (low)	C factor (high)
Asbestos-cement	140	140
Cast iron new	130	130
Cast iron 10 years	107	113
Cast iron 20 years	89	100
Cast iron 30 years	75	90
Cast iron 40 years	64	83
Cement-mortar lined ductile iron	140	140
Concrete	100	140
Copper	130	140
Steel	90	110
Galvanized iron	120	120
Polyethylene	140	140
Polyvinyl chloride (PVC)	150	150
Fiber-reinforced plastic (FRP)	150	150

Then,

$$\textit{Total dynamic head} = 7.5 + 31.1 + 14.91 = 53.51 \text{ m}$$

Assume the required system pressure is 27.5 m (from the available recommended range stated in Sect. 3.4). Following Eq. 3.11,

$$\begin{aligned}\textit{Pump head} &= \textit{Required system pressure} + \textit{Total dynamic head}\\ &= 27.5 + 53.51\\ &= 81.01 \text{ m}\end{aligned}$$

Thus, based on the determined pump head (i.e., 81.1 m) and the flow rate (i.e., 175 lpm), a pump selection should be on the available ‘pump curves.’

Rainwater harvesting systems provide an alternative water supply to meet consumer demand during short rainfall periods. This “make-up” or “back-up” system operates by a warning sign or switches to the alternate water supply. The make-up system requires high water quality if multiple water sources are available other than rainwater, municipal or private water. Integration of a rainwater harvesting system within a household plumbing system has shown in Fig. 3.12. The desired usages are toilet flushing, washing machine, irrigation system, and garden hose tap along with municipally supplied water. In this connection, regular water quality monitoring makes the system sustainable for a water smart city.

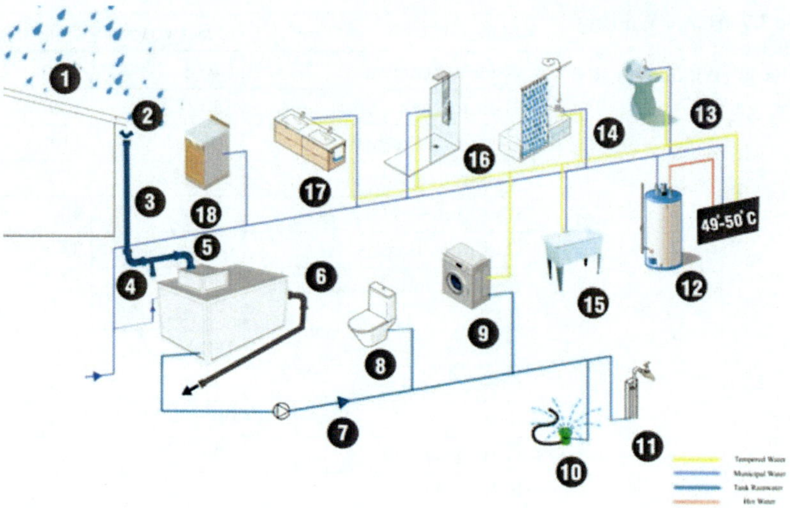

1: Roof, 2: Gutter, 3: Downspout, 4: First-flush diverter, 5: Rainwater tank with municipal water top-up, 6: Tank overflow pipe connected to the stormwater drainage system, 7: Pump, 8: Toilet; 9: Washing machine 10: Irrigation system 11: Garden hose tap, 12: Hot water system, 13: Hand basin, 14: Bath, 15: Laundry, 16: Shower, 17: Kitchen sinks, and 18: Dishwasher

Fig. 3.12 Integration of rainwater harvesting in a household plumbing system

3.6 Rainwater Quality

The catchment surface, storage material, environmental and financial conditions of the user, and rainwater overflow contaminate the harvested rainwater. In an urban area, anthropogenic sources of air pollution due to industry and significant roadways are threats to rainwater quality before the collection. On the catchment, overhanging foliage deposits leaves, pollen, and animal droppings, including birds, rodents, and squirrels, should be avoided. Catchment contaminates surface runoff during the conveyance either by the washing off the recently contaminated surface within rainfall events or the chemicals and metals leaching from the catchment material and rainwater storage tank material(s) or associated components within the tank. If overflows from the rainwater storage tank are directed to a city's storm sewer or an on-site soakaway pit, polluted backflow into the tank might happen during extreme storms. Therefore, source wise the harvested rainwater could be contaminated by:

Debris from the atmosphere (dust and dirt), foliage (leaves and branches), birds and animals (droppings), etc., reduce the aesthetic quality of the water and cause chemical and biological contaminations. Leaves and dust add chemical contamination with herbicides and pesticides. Biological contaminations, i.e., microscopic parasites, bacteria, and viruses, happen within the water due to bird and animal droppings.

Chemicals airborne chemicals intrude during rainwater collection. Volatile organic chemicals (VOCs) exist in plastics, glues, solvents, gasoline, greases, and oils. VOC contaminations take place through raindrops pass an atmosphere holding gasoline or solvent vapours. VOC contamination also spreads through improper construction practices in any part of the rainwater harvesting system. Synthetic Organic Chemicals (SOCs) contaminants introduce through dust, leaves, pesticides, herbicides, and similar human-made products. Compared to VOC contaminations, SOC contamination results from environmental exposure than poor construction issues.

Minerals are primarily inorganic salts, i.e., calcium carbonate, sodium bicarbonate, magnesium sulfate, and sodium chloride, which change water taste. Usually, these minerals intrude from the environment. Silica salts, often used by manufacturers for various products, could release long-term health-threatening asbestos in water.

Microbiological contaminants are present in harvested and stored rainwater in both pathogenic and non-pathogenic forms. Nonpathogenic microbes include protozoa, algae, bacteria, and viruses, reduce the aesthetic quality, require treatment facilities, and have higher operational and maintenance requirements. Thus, the concerned water quality parameters are pH, E-coli, fecal coliform, and chlorine. Similarly, open space collections would consider other parameters, including ammonia, chlorine, aluminum, turbidity, nitrate, and nitrite.

Sampling from a rainwater harvesting system should consider the following issues, i.e.:

- Hygienic handling and details knowledge on the desired water quality;
- Records on rainfall duration are needed as the runoff events influence the contaminant concentration. For example, contaminants in a rooftop rainwater harvesting system or higher pesticide concentrations in the agricultural field would dilute in the storage system and surface runoff tank, respectively;
- Appropriate sampling time and location should be identified based on regional meteorological and topographical knowledge.

Few or no relevant water quality tests might be available in-situ testing, i.e., field testing. Therefore, cares in sample handling is required. Thus, a rainwater quality database is available in a particular community. Water sampling and laboratory tests follow regional codes; for example, the American Public Health Association (APHA) developed standard water and wastewater handling (APHA 1998).

3.7 Rainwater Treatment

Selection of treatment units considers the desired water quality; treatment processes ensure water safety and aesthetic qualities. Location-wise treatment units are *Point-of-Entry* or POE to treat all of the water that enters the existing plumbing system and *Point-of-Use* or POU to treat water at the consumption point. POE treatment system recommends eliminating potential health threat contaminants; however, it lessens operational and maintenance requirements in the long run. POU treatment system adopted to improve the aesthetic quality of harvested water.

Pre-storage treatment devices incorporate with the conveyance network, and the treatment process operates by gravitational action. On the contrary, *post-storage treatment* devices require pressurized flow and energy. For rainwater harvesting systems, the treatment devices selection depends on:

- Permissible water quality requirements as per regional and national codes
- Applications of the harvested rainwater
- Locally available treatment devices or materials
- Disposal plan for the treatment process generated waste stream
- Operation and maintenance policy for the treatment.

3.7.1 Pre-Storage Treatment Devices

Pre-storage treatment devices include pre-filtration, first flush diversion, or settling. *Pre-filtration devices* located in the conveyance system can remove contaminants that might collect on the roof before the storage tank. Depending on the number of dry days between rainfalls, seasonal variation, land use (agricultural or industrial activities) for a specific area, the location of freeways, and the presence of overhanging trees would influence the nature of contaminants. Usually, the contaminants include leaves, twigs, atmospheric dust, pollens, pesticide residues, insects, birds/and animal droppings. The most commonly used pre-filtration components are:

- Gutter guards

They are attached over the top of gutters to prevent entering debris and blocking the gutters (Fig. 3.13a).

- Downspout filters are also known as leaf and/debris diversion rain heads

Due to the absence of gutter guards, water flows from a gutter could carry contaminants. This diversion passes water over the angled screen to retain leaf/debris are, and the water passing continues through the screen. Commercial leaf diverters are also available for different rainfall intensity areas (Fig. 3.13b). A downspout filter can also adopt in addition to the gutter guards (Fig. 3.13c). It can be attached (i) to the top of the downspout, or (ii) any suitable location along the length of the downspout, or (iii) in the ground below the downspout (CMHC 2012).

The first-flush diversion installs to divert or flush away the first collected rainwater over the catchment surface before entering the storage. This flushing can be done manually or following the regional guidance (Fig. 3.14a). The suggested approximately 30 liters in Malaysia (Sehgal 2005) 40 liters in the US (TWDB 2005) of seasonal first rainwater recommends flush for every 100 m^2 of the catchment area. Generally, 0.5–1.5 mm of diversion height usually is used in Canada (CMHC 2012). A standpipe of known diversion height can be designed or installed according to

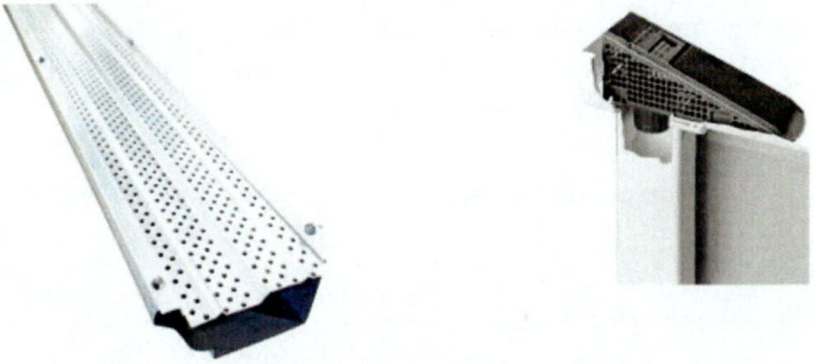

(a) Gutter guards prevent debris from clogging the gutters

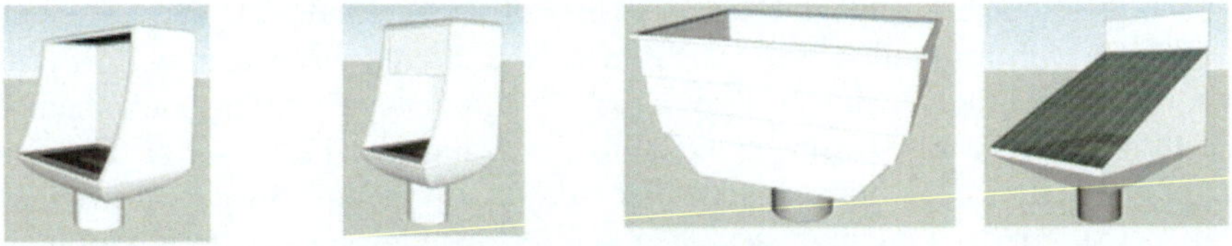

leaf beater - for low rainfall areas

leaf slider

leaf catcher - gutter or wall mounted

leaf eater - for high rainfall areas

(b) Leaf and/ debris diversion rain heads

(c) Downspout filters prevent debris from entering the tank

Fig. 3.13 Pre-filtration devices

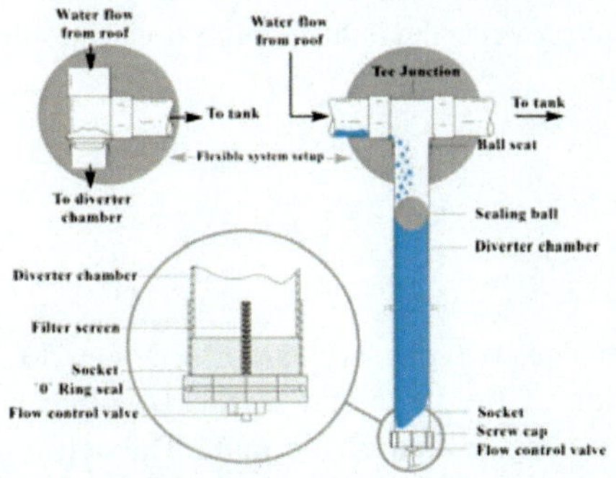

(a) First-flush device: during the diverter, chamber filling with contaminated water

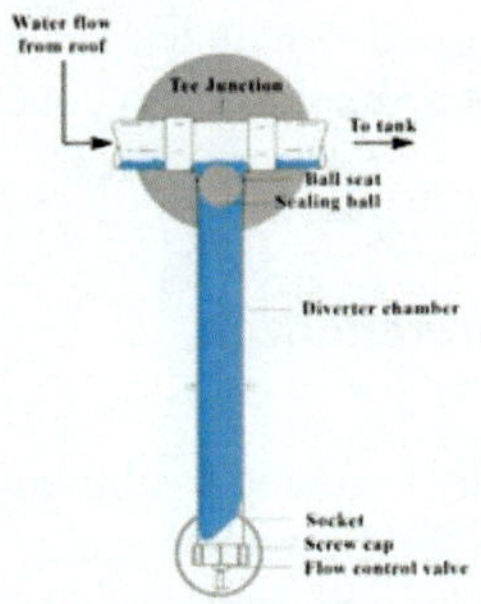

(b) First-flush device: while the diverter chamber is complete and the ball seals the chamber

Fig. 3.14 First-flush device

the downspout following the regional guidance. As shown in Fig. 3.14b, once the standpipe fills with contaminated rainwater, cleaner rainwater flows into the storage system. Thus, the diversion volume and chamber size are:

$$\text{Diversion volume (liter)} = \text{Diversion height (mm)} \times \text{Catchment area} \left(\text{m}^2\right) \tag{3.14}$$

$$\textit{Height of first} - \textit{flush chamber} \text{ (mm)} = \frac{4 \times \textit{Diversion volume} \text{ (liter)} \times 1000}{3.14 \times \left[\textit{Pipe diameter} \text{ (mm)}^2\right]} \tag{3.15}$$

The pre-storage treatment system consists of either a sediment tank or a sediment tank with a filtration process based on the desired water quality. Collected rainwater from the catchment allows passing through the settling tank/chamber where the debris can settle down and collect as sediment. The treated or 'clarified' water transferred to the storage tank either considers further treatment or distribution (Fig. 3.15). For drinking purposes, then the treatment advances with filtration and disinfection. Figure 3.15a shows a rainwater storage tank followed by a sedimentation tank. Figure 3.15b presents the pre-storage treatment facilities for the rainwater storage tank; the pre-storage treatment facilities include sedimentation, filtration, and disinfection.

3.7.2 Post-Storage Treatment Devices

Post-storage treatment devices include filtration, disinfection, and other required treatment to ensure the aesthetical properties of water. The conveyance system comprises post-storage treatment units in Fig. 3.8 within this chapter. Commonly, a 5-micron particle filtration and ultraviolet (UV) disinfection (Fig. 3.16). A reduced pressure valve is also included as the backflow prevents the reduced pressure principle to protect water supplies from contamination.

3.7.3 Sedimentation

Conventionally horizontal sedimentation involves two primary parameters, i.e., *flow velocity* with the flow over the cross-section of the tank; and *surface loading* or *settling velocity* while flow passes over the surface area of the tank. The settling velocity (v_s) of a particle is the most critical design parameter to compute the efficiency of discrete sedimentation.

Complete removal of suspended particles occurs in a sedimentation tank if the terminal settling velocity equals the overflow velocity (v_0), i.e., $v_{s.} = v_0$ (Fig. 3.17). Therefore, the percentage of removal is $100\frac{v_s}{v_0}$. Thus apply the famous Stokes' law

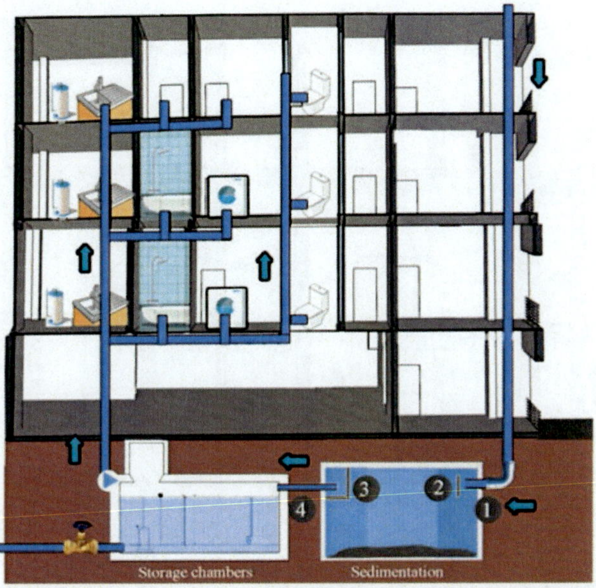

a: Integrated sedimentation and rainwater storage chambers

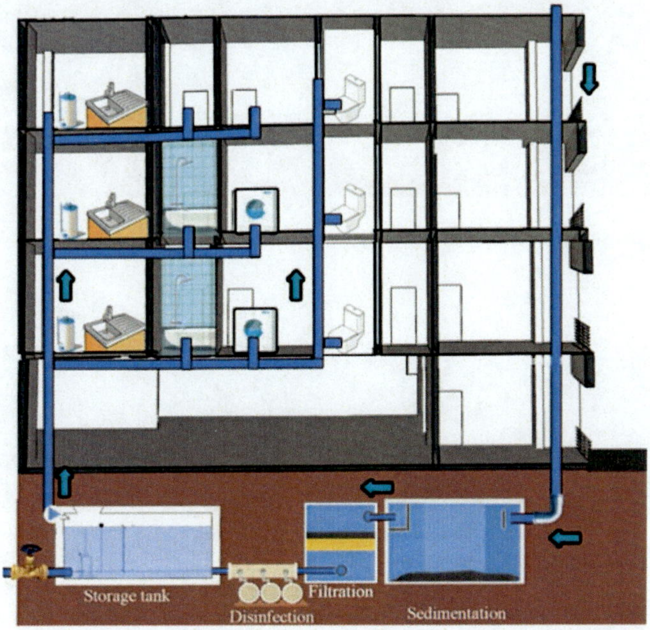

b: Pre storage treatment

Fig. 3.15 Pre-storage treatment system

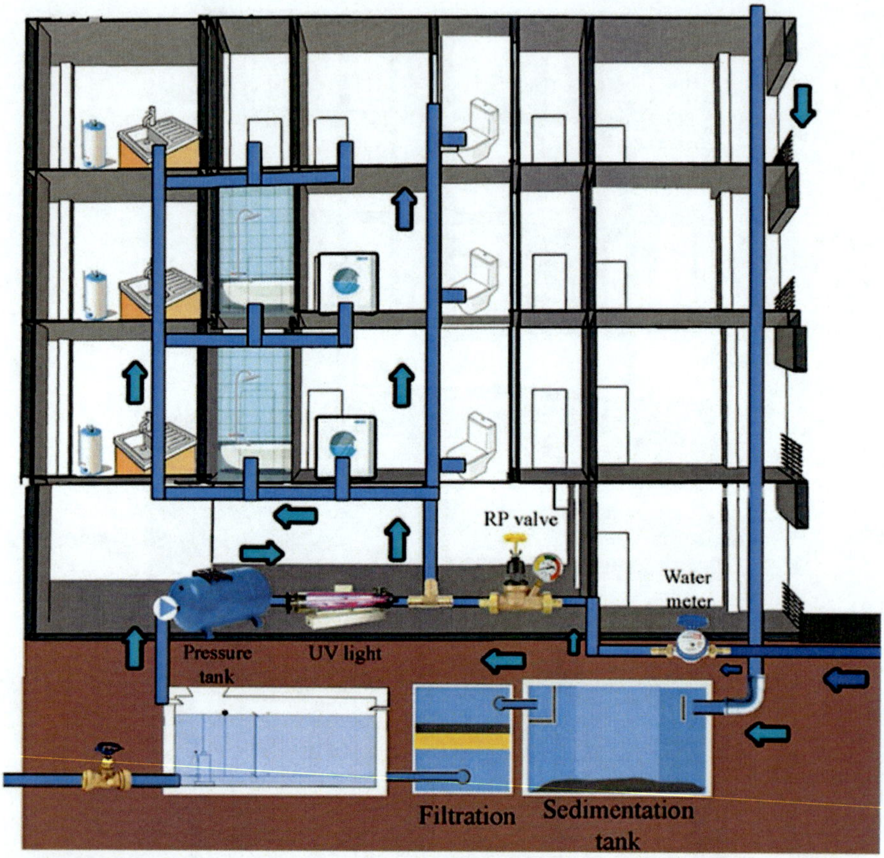

Fig. 3.16 Post storage treatment

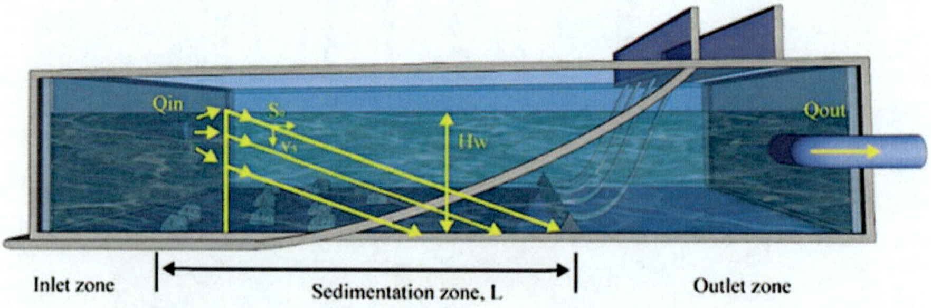

Fig. 3.17 Settling conditions in an ideal rectangular sedimentation tank

for laminar settling, depending on the water viscosity and the particle's size and density. There are two components of velocity while particles falling through the sedimentation tank, i.e.:

$$\text{Vertical component, } v_s = \frac{(\rho_p - \rho)gd^2}{18\mu} \quad (3.16)$$

$$\text{Horizontal component, } v_h = \frac{Q}{A} \quad (3.17)$$

The particle path is the vector sum of these two velocities. On the other hand, $v_0 = \frac{Q_0}{A}$. Usually, these tanks settle solids and remove the floating materials, reducing the load on the biological treatment units. An efficient sedimentation tank should settle 50–65% of the suspended solids and reduce 25–40% of the Biochemical Oxygen Demand (BOD).

Design details of a sedimentation tank:

- Detention period: 3–4 h (plain sedimentation), and 2–2.5 h (coagulated sedimentation)
- Velocity of flow: within 0.30 m/min (horizontal flow).
- Tank dimensions: for a *rectangular tank,* the length is 30–100 m, and the width is 6 m–10 m. A recommended length–width ratio L: B is 3 to 5:1.

For a *circular tank,* the diameter ranges between 20 and 40 m and should be within 60 m.

- Depth ranges between 2.5 and 5.0 m.
- Surface overflow rate: 12,000 to 18,000 liter/d/m^2 tank area (plain sedimentation); 24,000–30,000 liter/d/m^2 tank area (thoroughly flocculated water).
- Slopes towards the inlet: 1% (rectangular tank); 8% (circular tank).

Example problem 3.5 Design a rectangular sedimentation tank to treat raw water of 230 m^3/day. The expected detention period is 3 h.

Solution: Raw water flow is of 230 m^3/day. Detention period is 3 h.

Volume of tank = Flow × Detention period = $230 \times \frac{3}{24} = 28.75$ m^3.

Assume depth of tank = 1.5 m.

Surface area = 28.75/1.5 = 19.17 m^2.

If, L/B = 3 (assumed). Then,

$3B^2 = 19.17$ m^2 i.e. B = 2.53 m.

L = 3B = 2.53 × 3 = 7.59 m.

Hence, surface loading (Overflow rate) = $\frac{(230)}{19.17}$ = 11.998 m^3/d/m^2 < 18.000 m^3/d/m^2 (ok).

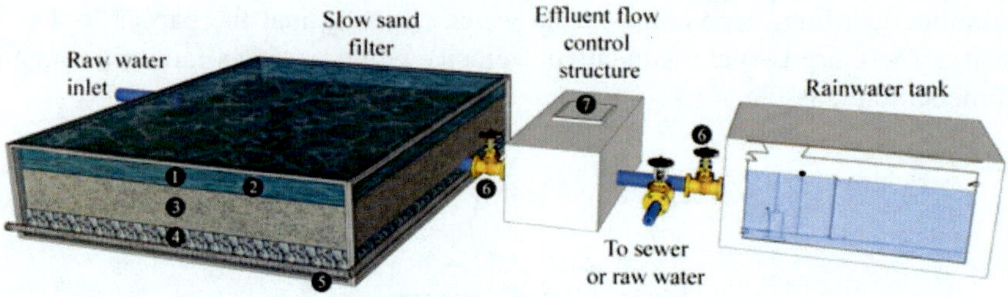

1: Supernatant water reservoir, 2: Schmutzdecke, 3: Sand bed, 4: Supporting gravel, 5: Underdrainage, 6: Control valve, 7: Manhole

Fig. 3.18 Typical Slow Sand Filter (SSF)

3.7.4 Filtration

Gravity filtration within a concrete tank is standard for city water plants. There is also pressure filtration and involvements of hi-tech filtration viz. activated carbon filter, membrane filtration. Slow Sand Filter (SSF) acts as centralized and semi-centralized water purification systems based on gravity flow. The influent water passes the sand-gravel bed to remove turbidity and pathogenic organisms and obtain treated water from the underdrain system. SSF system is a single treatment step, comprises sedimentation, straining, adsorption, chemical, and bacteriological. SSF is the most desirable due to its simple construction and easy operation and maintenance. There are three major components in the SSF system: filter, effluent flow control structure, and treated rainwater tank (Fig. 3.18).

The water purification process comprises mechanical and physical–chemical methods. The resulting sediment and organic matter accumulation form a thin layer on the sand surface, known as 'schmutzdecke'. SSF is suitable for treating waters with lower contents of colours, turbidity, and bacterial components. Thus, SSF removes significant coliform (up to 95%), cryptosporidium and Giardia cysts (up to 99%), moderate colour (up to 75%), and few total organic contents (10%). For a typical sedimentation tank design, details are (Huisman and Wood 1974; Galvis et al. 1998):

- The purification mechanisms extend with a bed depth of 0.8 m–1.2 m;
- Effective media size: ranges between 0.15–0.45 mm and uniformity coefficient (Cu)[1] between 1.8 and 3.0;
- Filtration rate: 0.08–0.14 m/h; and
- Supernatant waters: 0.9–1.5 m.

Rapid Sand Filter (RSF) follows a similar operational technique to the SSF for water filtration. The influent water in an RSF is already relatively clear and operates quicker than the SSFs. The significant parts of gravity RSF are the water reservoir, filter media, perforated laterals, drains, and filtered water for the rain tank (Fig. 3.19).

[1] Uniformity coefficient (Cu) is the ratio of D60 to D10.

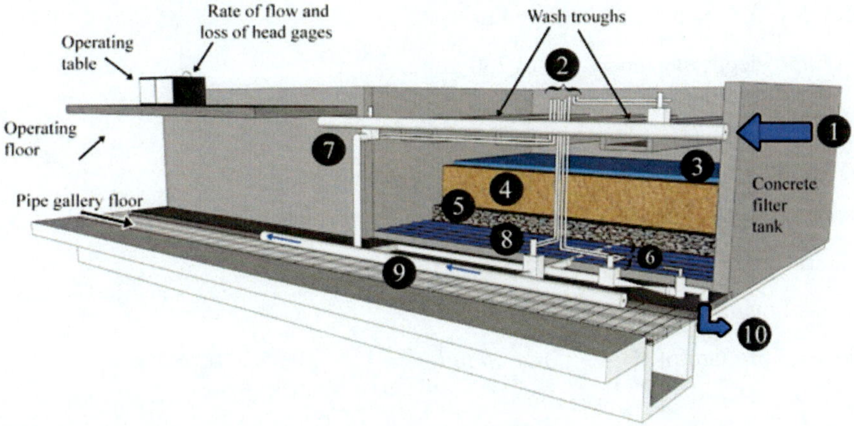

1. Influent to filters, 2. Pressure lines to hydraulic lines to hydraulic valves from operating tables, 3. Water reservoir, 4. Filter sand, 5. Graded gravel; 6. Perforated laterals, 7. Filter drain, 8. Filter to waste, 9. Wash line, 10. Effluent to rain tank

Fig. 3.19 Typical rapid sand filter (RSF)

Influent passed to filter bed, and adequate media size is within 0.4–0.7 mm with the uniformity coefficient, Cu < 1.5. Underneath the filter bed, a perforated lateral transfers the wastewater through the washing line. Most RSFs contain a control system to regulate water flow rates through the filter. The expected filtration rate is 4–21 m/h per m^2. As a part of regular filter maintenances, dedicated components include valves, *Loss of Head* (LoH) indicator, surface washers, and a backwash pump. Thus, RSF removes significant coliform (up to 90%), cryptosporidium and Giardia cysts (50–90%), few colour (up to 10%), and total organic content (5%). A *pressure filter* contains a closed water-tight cylindrical drum based on the similar mechanism of RSF, and filtered water passes over the sand bed due to applied pressure.

Example problem 3.6 Design an RSF to treat 230 m^3/day, allowing 0.5% filtered water backwashing. The duration allowance for the backwashing is 30 min a day. Assume additional data (if needed).

Solution:

Total filtered water 23.5 h $= \frac{230+\left(230\times\frac{0.5}{100}\right)}{23.5} = 9.84$ m/h.

Let, filtration rate is 0.15 m/h/m^2 of bed.

Area of filter $= \frac{231.15}{23.5} \times \frac{1}{0.15} = 65.57\ m^2$.

Consider two units. Bed area for an individual unit $= \frac{65.57\ m^2}{2} = 32.78\ m^2$.

Considering, aspect ratio of the proposed tank is 1.3. i.e. L/B = 1.3; $1.3\ B^2 = 32.78\ m^2$.

B = 5.1 m; L = 5.1 × 1.3 = 6.63 m.

Assume, depth of sand = 500 to 750 mm.

Drainage system

Typically, the total area of holes is 0.2–0.5% of the bed area.

Area of lateral holes = $\frac{0.2}{100} \times 32.78 = 0.0655\ m^2$ [Assume 0.2% of bed area].

Lateral area = 2 (Area of lateral holes)

Area of manifold = 2 × Lateral area

So, area of manifold = $4 \times area\ of\ holes = 4 \times 0.0655 = 0.262\ \text{m}^2$

$$Then,\ diameter\ of\ manifold = \left(4 \times \frac{0.262}{\pi}\right)^{1/2} = 577\ \text{mm}$$

Assume, c/c of lateral = 300 mm. Total lateral numbers over L = $\frac{6.63 \times 1000}{300} = 22.1 \approx 22$ on either side.

Length of lateral = $\frac{5.1}{2} - \frac{0.577}{2} = 2.26\,\text{m}$

The cross-sectional area of lateral = $2 \times area\ of\ perforations\ per\ lateral.$

Assume the diameter of holes = 13 mm, then the number of holes:

$$n\frac{\pi(13)^2}{4} = 0.0655 \times 10^6$$

$$\therefore n = \frac{4 \times 0.0655 \times 10^6}{\pi(13)^2} = 493.47 \approx 494$$

As 22 laterals are on either side. Then, number of holes per lateral = $\frac{494}{44} = 11.22 \approx 11$.

Area of perforations per lateral = $11 \times \frac{\pi(13)^2}{4} = 1460\ \text{mm}^2$.

Spacing of holes = $\frac{1460}{11} = 132.7$ mm

$$\begin{aligned}\text{The cross - sectional area of lateral} &= 2 \times Area\ of\ perforations\ per\ lateral\\ &= 2 \times 1460 = 2920\ \text{mm}^2\end{aligned}$$

Diameter of lateral = $\sqrt{\frac{(4 \times 2920)}{\pi}} = 60.97 \approx 61$ mm.

Check: Length of lateral, $l < 60$ d.

$= 60 \times 61 = 3.66$ m. $l = 2.26$ m (Hence acceptable).

Assume, rising wash water velocity in the bed is 500 mm/min.

Wash water discharge per bed = $\frac{0.5}{60} \times 5.1 \times 6.63 = 0.282\ m^3/s$.

The velocity of flow through lateral = $\frac{0.282}{Total\ lateral\ area} = \frac{0.282 \times 10^6}{44 \times 2920} = 2.19$ m/s (ok).

Manifold velocity = $\frac{0.282}{0.262} = 1.08\frac{m}{s} < 2.25\frac{m}{s}$ (ok).

Washwater gutter

Discharge of washwater per bed = $0.282\ m^3/s$. Size of bed = 5.1 m × 6.63 m.

Assume three troughs running lengthwise at $\frac{5.1}{3} = 1.7$ m c/c.

Discharge per trough = Q/3 = $\frac{0.282}{3} = 0.094\ m^3/s$.

$$Q = 1.71 \times b \times h^{3/2}$$

Assume b = 0.3 m

$$h^{3/2} = \frac{0.094}{1.71 \times 0.3} = 0.183$$

then, h = 0.326 m = 326 mm ≈ 350 mm = 350 + (free board) 50 mm = 400 mm; slope 1 in 40.

Clearwater reservoir for backwashing

For four hours filter capacity, the capacity of tank = $\frac{4 \times 5000 \times 5.1 \times 6.63 \times 2}{1000} = 1352.52\ m^3$.

Assume depth d = 5 m. Surface area = $\frac{1352.52}{5} = 270.5\ m^2$.

L/B = 2; $2B^2 = 270.5$; B = 12 m, and L = 24 m.

Inlet pipe diameter coming from two filter = 500 mm.

Velocity < 0.6 m/s. Washwater pipe diameter to overhead tank = 675 mm.

Assume, Air compressor unit = 1000 liters of air/min/m^2 bed area.

For 5 min, air required = 1000 × 5.1 × 6.63 × 5 × 2 = 338 $m^3$of air.

Membrane filtration or cross-flow filtration is a gentle, physical separation process. In membrane filtration, the filtrate is 'permeate,' the concentrate is 'retentate', and the build-up of dry sludge during membrane filtration is the 'coating'. Membrane filtration applies in addition to chemicals to ensure aesthetical issues. The range of membrane filtration processes has summarized in Fig. 3.20. These are microfiltration (MF), ultrafiltration (UF), nanofiltration (NF), and reverse osmosis (RO). *MF membranes* comprise the largest pore size (0.1–10 μM) to remove large particles and various microorganisms; yeast, asbestos, bacteria, atmospheric dust. *UF membranes* contain smaller pores (0.01–0.10 μM) than MF membranes, and therefore, UF membranes remove atmospheric dust, viruses, pathogen endotoxin, bacteria, and soluble macromolecules, i.e., proteins. Relatively new *NF membranes* or "loose" reverse osmosis membranes are porous membranes on the order of ten angstroms or

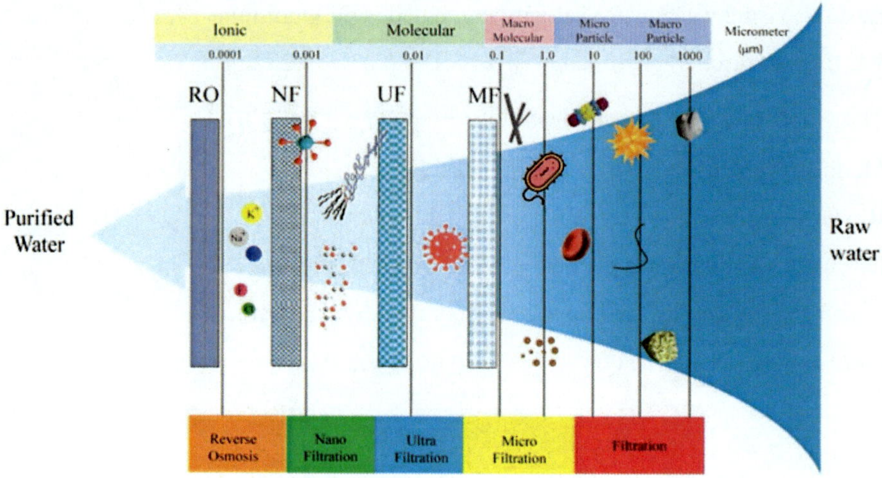

Fig. 3.20 Membrane filtration

less. NF membranes exhibit performance between RO and UF membranes. Therefore, RO membranes are efficiently non-porous to exclude particles with low molar mass species, viz. salt ions, organics, etc.

3.7.5 Disinfection

For harvested rainwater as a potable water source, disinfection applies to remove or inactivates micro-organisms using chemical or physical treatment. The corresponding methods for physical disinfectants include thermal treatment, Ultraviolet light (UV), and Ozone. UV reactor can be placed after filtration or in a combination of ozone applications. While the treatment process includes ozone and UV disinfection systems, the ozone treats the bulk of the water in the collection tank, and UV uses to treat after filtration (Fig. 3.21). Chlorination is the chemical disinfection process aimed to kill the remaining pathogens by adding residual chlorine.

WHO guidelines permit the free residual chlorine concentration in drinking water between 0.2 and 0.5 mg/liter, and the maximum allowable chlorine concentration is 5 mg/liter (WHO 2014). Chlorine exists in bleaching powder (i.e., Calcium Hypochlorite: $CaOCl_2$) and commercial bleach solution (i.e., Sodium Hypochlorite: NaClO). Chlorine concentration is approximately 5 mg/liters produce from the 40 mL of liquid sodium hypochlorite, i.e., 12.5% available chlorine per 1000 liters of water. On the other hand, powdered calcium hypochlorite (75% available chlorine) dose is 7000 mg/m^3 water.

Example problem 3.7 For the seven-storied building, stated in *Example problem 3.3*, the available harvestable water was found as 3.95 m^3/day. A commercial

Fig. 3.21 Ultraviolet light (UV)

bleaching powder with 65–70% available chlorine is planning to apply for disinfection. Determine the required bleaching powder to maintain 2.5 mg/liter chlorine in the disinfected water?

Solution:

Total Cl_2 demand $= 2.5 + 0.2$ as residual Cl_2.

Therefore, per liter of water required 2.7 mg/liter.

Then, 3.95×10^3 liter requires $= 3.95 \times 10^3 \times 2.7 \frac{\text{mg}}{Cl_2} = 10.665 \text{ kg } of \ Cl_2$.

Available Cl_2 is 70%, i.e., 1 kg of Cl_2 is available from 1.43 kg of powder. Then the required bleaching powder $= 1.43 \times 10.665 = 15.25$ kg.

3.8 Design and Installation Procedures

The design procedure of rainwater harvesting comprises of following steps:

Step 1: Collection.

- Rooftop/open surfaces, those have complied with regional catchment material standards, are usually recommended for collecting rainwater;
- The catchment surface needs to be as large as possible to maximize the volume of collected rainwater;
- Catchment material should be with negligible collection losses, viz. steel;
- Collection of rainwater from green roofs and roofs with overhanging foliage is not recommended.

Step 2: Conveyance.

Plan the layout of the conveyance network, determine the location of the rainwater tank (above or below ground) and then decide the drainage facility from the downspout(s) to the tank. For newly buried conveyance drainage pipes, special care needs for underground utility services, i.e., gas, electricity, water, stormwater, wastewater, phone, or cable lines. The following issues also should be taken into consideration:

- Aluminum or galvanized steel are recommended as gutter and downspout materials, while copper, wood, vinyl, and plastic are not recommended;
- Custom-fabricated gutters installation avoids seams along the gutter length;
- The gutter should be slope towards rainwater storage tank direction, and a minimum pitch of 0.5–2% should be maintained throughout the gutter length;
- The required gutter size for a given roof drainage area should follow territorial codes and regulations; if not available, use Eq. 3.1;
- The downspout(s) should be located near the rainwater storage tank but not inside the building corners. The number of downspouts can be defined using Eq. 3.2;
- Downspout offsets should be within 3.0 m;
- The selected pipe materials need to be approved by regional codes and industry standards. Acrylonitrile Butadiene Styrene (ABS) and Poly Vinyl Chloride (PVC) pipe materials are recommended. A minimum slope of 0.5–2% maintains throughout the pipe length. For cold weather, insulation or heat tracing needs to be available as pipe-freezing protections.

Step 3: Rainwater storage.

Components installed in the rainwater storage tank include a pump intake, sensors, control equipment, and electrical wiring for internal connections. Each pipe connection and electrical connection should be appropriately sealed and watertight. The associated design considerations are:

- Rainwater collection pipe should enter the storage tank maintaining a height above (preferably 50 mm) the bottom of the overflow drain;
- Determine the location of the rainwater storage tank considering its placement, drainage facilities, and accessibility. For cold weather, tanks at risk for freezing should be protected by winterizing (insulation, heat tracing) or decommissioning (through a tank bypass or tank drain valve);
- Determine the rainwater storage tank capacity to meet the design demand allowing 20% of tank capacity as unused volume or 'dead space' and conveyance losses from treatment devices;
- If a make-up system exists, determine the mode of the make-up system automatically or manually associated with the existing rain storage. Then, plan for the top-up system and appropriate water level sensors are required, electrical conduit, and rainwater service conduit installed accordingly.
- Material selection for the rainwater storage tank considering its position (overhead or underground or integrated), required storage volume, technical specifications, and connected rainwater fixtures and desired quality;

- Tank access openings should be as per the territorial required standard; otherwise, at least 450 mm. The entry of small animals or insects into the rainwater storage tank should be strictly restricted.

Step 4: Water treatment.

Factors that influence the quality of rainwater in the rainwater harvesting system should be identified and mitigated through proper design and installation:

- Catchment surfaces are subjected to contamination risks due to surface material, surface positioning, and the proximity to sources of air pollution. Chemicals and metals leach from surface material, kitchen cooktop vent, and dryer vent might add grease and lint, overhanging foliages, and animal droppings are possible sources of physical and biological contamination of the collected rainwater. To avoid poor rainwater quality, restrain the collection of runoff from the catchment area's risky sections.
- Conveyance networks with poorly sealed joints could be at risk of leaching insufficient quality groundwater/surface water or even entry of animals and insects. Underground pipe connections and fittings should be secure to overcome this issue.
- Rainwater storage tank pollutes settled sediments, insects, rodents or debris, algal growth, chemicals, and metals leaching from tank material. Proper tank material selection, tank covering, reduction in leakage within the conveyance system, and pump placement are required.
- Determine rainwater quality and treatment requirements as per consumers' choice. Usually, for non-potable water, a pre-storage treatment device is required. Filtration, disinfection, pH adjustment for acidic rainwater needs to meet drinking water standards.
- Pre-storage treatment devices should be capable of handling the peak runoff from the catchment surface.

Step 5: Treated water storage.

The settling tank or settling chamber size should follow the interim storage of a prescribed volume of runoff.

Pre-storage treatment filtration devices

- Filtering system strengthened including high-quality gutter guards, leaf screens placed on the downspout, rainwater filter installed either with conveyance pipe network or within the tank;
- Estimate initial and continuous collection loss factors;
- Pre-storage treatment installed devices should be readily accessible.

Post-storage treatment devices (if applicable)

- Post-storage treatment devices require less involvement if there is a provision for Pre-storage treatment;

- Post-storage treatment devices follow the maximum flow rate of the pressure system;
- Post-storage treatment installed devices should be readily accessible.

Step 6: Water distribution.

Treated water is delivered by a pump within the building using a pressurized bladder system.

- Determine the fixtures connected to rainwater;
- Pump selection should consider style and operating characteristics, required flow rate, pump head (stated in Eq. 3.11), and collect the 'pump curve' charts;
- If a pressure tank requires, this should be selected considering pump controller configuration and pump flow rate;
- Plan route of the (a) rainwater service pipe; (b) electrical conduits; and (c) underground utility service lines.
- Rainwater service pipes should be of the standard pipe materials, pipe size, and tank connection. This piping should maintain a gap of around 50 mm between the inlet pipe and the overflow drainage piping. Similarly, rainwater supply pipes are selected based on the permitted pipe materials and pipe size. Estimation of these pipes could be using Eqs. 3.12 and 3.13.
- Rainwater piping should be supported and protected as per regional standards and guidelines for considering their surrounding consequences.
- Pipe markings using approved texts, colour, and spacing.

3.9 Management Strategies

Step 1: Collection.

- The catchment surface inspection needs once every six months to ensure cleanliness and before starting the rainy season after the dry period;
- Alternative approaches adopt to secure the collections during the presence of ice and winds;
- All necessary safety precautions should follow during the inspection, cleaning, or repairing of the catchment surface and parts of the conveyance network.

Step 2: Conveyance.

Plan the layout of the conveyance network, determine the location of the rainwater tank. Then,

- The gutters and downspouts inspection should be conducted once every six months to remove accumulated dirt/debris and remove/replace damaged components with the collection system;
- Identification of faulty pumps, water level sensors, or other control equipment is required at least once a year.

Step 3: Rainwater storage.

- For both below-ground storage and above-ground water tanks, leaks may be inspected at least once a year through the poor performance of the rainwater supply;
- Sediment/debris accumulation at least once a year from the storage tank and, depends on the treatment provided, appear at the POU;
- Ensure health and safety while inspecting, cleaning, or repairing the tank;
- Inspect the make-up system once per six months. If the makeup system does not operate, perform the float switch, solenoid valve, and top-up drainage pipe is required.

Step 4: Water treatment.

Consider site-specific risk factors to select appropriate maintenance schemes to mitigate the rainwater quality risks. Generally, these schemes include:

- Pre-storage treatment devices inspection should be at least twice a year or more,
- Care requires to remove dirt and debris blocking flow through the filter;
- Post-storage treatment devices check-up should be once every three months.

Step 5: Treated water storage.

Size selection of the settling tank or settling chamber considers designed storage volume. Pre-storage treatment filtration devices.

i. Filtering system includes high-quality gutter guards, leaf screens placed on the downspout, and the filter installed either in conveyance pipe network or within the tank;
ii. Estimate both initial and continuous loss factors;
iii. Pre-storage treatment devices installation ensures their readily accessible.

Post-storage treatment devices (where applicable).

i. Pre-storage treatment facilities reduce treatment action by the post-storage treatment devices;
ii. Selection of the post-storage treatment devices depends on the maximum flow rate of the pressure system;
iii. Readily accessible post-storage treatment devices installed.

Step 6: Water distribution.

The stored rainwater is often used for non-potable use, and signboards alert the consumers associated with the labeled or different colour pipes (Fig. 3.22:). The pump and pressurized distribution system need to operate correctly, and inspection once annually is recommended.

- The pump cycles repeatedly interrupt the improper pressure system; it might be an issue with the pressure tank.
- Leakage inspection of the pressure sensor/switch, pipes, and shut-off valves, other control valves should be at once while rainwater supply is not adequate.

Fig. 3.22 Rainwater warning sign

References

ABCB (2019a) Cross-connection control. Australian Building Codes Board (ABCB).

ABCB (2019b) National construction code volume three—plumbing code of Australia, Australian Building Codes Board (ABCB), Canberra, ACT.

APHA (1998) Standard methods for the examination of water and wastewater, 20th edn. American Public Health Association, American Water Works Association and Water Environmental Federation, Washington DC.

AS/NZS (2018) AS/NZS 3500 plumbing and drainage (series).

ASCE and WFCP (1969) Design and construction of sanitary storm sewers. American society of civil engineers and water pollution control federation, ASCE manuals and reports on engineering practice no. 37 and WPCF manual of practice no. 9.

BNBC (2020) *Bangladesh National Building Code*. Ministry of Housing and Public Works.

BSI (2018) On-site non-potable water systems. Systems for the use of rainwater. BS EN 16941-1:2018. 978 0 580 89028 4.

CMHC (2012) Guidelines for residential rainwater harvesting systems handbook. Canada Mortgage and Housing Corporation (CMHC).

CSA (2006) Design and installation of non-potable water systems/maintenance and field testing of non-potable water systems

Despins C (2012) Guidelines for residential rainwater harvesting systems handbook. ISBN 978-1-100-21183-1

Fewkes A (2000) Modelling the performance of rainwater collection systems: towards a generalised approach. Urban Water 1(4):323–333. https://doi.org/10.1016/S1462-0758(00)00026-1

Galvis G et al (1998) Multi-stage filtration and innovation water treatment technology. International reference centre for community water supply and sanitation. The Hague, Netherlands and Universidad del valle instituto CINARA, Cali, Colombia.

Ghisi E et al (2009) Rainwater harvesting in petrol stations in Brasília: potential for potable water savings and investment feasibility analysis. Resour Conserv Recycl 54(2):79–85. https://doi.org/10.1016/j.resconrec.2009.06.010

Haq SA (2016) Harvesting rainwater from buildings. Springer.

Hofstein A et al (2004) Providing high school chemistry students with opportunities to develop learning skills in an inquiry-type laboratory: a case stud. Research report. Int J Sci Educ 26(1):47–62. https://doi.org/10.1080/0950069032000070342

Huisman L, Wood W (1974) Slow sand filtration. World Health Organization, Geneva, Switzerland. In: Juhna T, Melin E (eds) 2006. Ozonation and biofiltration in water treatment—operational status and optimization issues. Techneau (WP5.3 operation of water treatment facilities)

IAPMO and NSF (2013) ARCSA/ASPE/ANSI 63–2013: rainwater catchment systems.

Indian Standard (2008) Roof top rainwater harvesting—guidelines [WRD 3: ground water and related investigations]. IS 15797 : 2008

Jenkins D et al (1978) Feasibility of rainwater collection systems in California, Contribution No. 173, Californian Water Resources Centre, University of California

Khoo TC (2009) Singapore water: yesterday, today and tomorrow BT—water management in 2020 and beyond. In: Biswas AK, Tortajada C, Izquierdo R (eds) Berlin, Heidelberg: Springer Berlin Heidelberg, pp 237–250. https://doi.org/10.1007/978-3-540-89346-2_12

Lancaster B (2006) Guiding principles to welcome rain into your life and landscape. In: Rainwater harvesting for drylands and beyond, vol 1. Rainsource Press, Tucson, Arizonas.

Liaw C, Tsai Y (2004) Optimum storage volume of rooftop rain water harvesting systems for domestic use. J Am Water Resour Assoc 40:901–912

McCuen RH (2016) Hydrologic analysis and design, 4th. Pearson Education

Nagy IV et al (2002) Application of the moran model in reservoir storage design. In: Hydrological dimensioning and operation of reservoirs. Water Sci and Tech Library, vol 39. Springer, Dordrecht. https://doi.org/10.1007/978-94-015-9894-1_10

Nath K et al (2006) Household water storage, handling and point-of-use treatment. Int Sci Forum Home Hyg (IFH).

NRC (2015) National building code of Canada, 14th edn. National Research Council Canada, Ottawa.

NSF (2016) NSF P151: Certification of rainwater catchment system components. www.nsf.org

Pacey A, Cullis A (1989) Rainwater harvesting: the collection of rainfall and runoff in rural areas. Intermediate Technology Publications, London.

Schiller EJ, Latham BG (1987) A comparison of commonly used hydrologic design methods for rainwater collectors. Int J Water Resour Dev 3(3):165–170, Routledge. https://doi.org/10.1080/07900628708722345

Sehgal JD (2005) A guide to rainwater harvesting in Malaysia. Rotary Club of Johor Bahru, District 3319 of Rotary International.

TWDB (2005) The Texas manual on rainwater harvesting system, 3rd edn. Texas Water Development Board, Austin, Texas. http://www.ecy.wa.gov

Vickers M (2001) Handbook of water use and conservation. WaterPlow Press, Amherst (MA), pp 12–207.

Ward S et al (2010) Harvested rainwater quality: the importance of appropriate design. Water Sci and Tech 61(7):1707–1714. https://doi.org/10.2166/wst.2010.102

WHO (2014) Water Safety in Distribution Systems: Public Health, Environmental and Social Determinants Water, Sanitation, Hygiene and Health. WHO/FWC/WSH/14.03. WHO Document Production Services, Geneva, Switzerland

Williams GS, Hazen A (1920) Hydraulic tables: the elements of gagings and the friction of water flowing in pipes, aqueducts, sewers, etc., as determined by the Hazen and Williams formula and the flow of water over sharp-edged and irregular weirs.
Worm J, Hattum T (2006) Rainwater harvesting for domestic use, 1st edn. Agromisa Foundation and CTA, Wageningen, The Netherlands.

Chapter 4
Stormwater Management

4.1 Introduction

Urbanized area is a driver of stormwater flood risk. The rainwater harvesting system has been considered as a part of the stormwater management system. On the other hand, handling overflows from rainwater storages is usually followed by stormwater management requirements. In the twenty-first century, stormwater management is the lifeline in sustaining adjacent stream ecosystem services and community resilience. This chapter solely deals with rain-induced stormwater management to facilitate the water smart city. Characterization of urban stormwater runoff, this chapter presents the historical evolution of their management, following current trends, technological advancement, codes, worldwide standards, design and installment consideration, management strategies, and the relevant worked-out examples. Conventionally, stormwater is solely managed by urban water and sanitation experts; fast stormwater disposal outside the cities remains the top priority. Artificialisation of soils and alteration of the hydrological cycle results in urban floods, urban waterlogging, lowering aquifer recharge, surface water pollution, and adverse impacts on ecology, etc. Due to urbanization, it emerged to build a water smart city that emphasized integrated sustainable approaches. Thus, the target development on the stormwater management techniques worldwide embraces interaction with the ecology, biodiversity, bio-inspiration, architectural design, landscape and water values planning, urban water resources, natives' well-being, and socio-economic aspects. Few developed countries are already practicing these techniques and reasonably address problems engendered by the traditional approaches. Description of these techniques, their design and installation consideration, management strategies include. For this chapter, work-out examples are designed based on the existing practices/learnings.

© The Author(s), under exclusive license to Springer Nature Switzerland AG 2022

A. Akter, *Rainwater Harvesting—Building a Water Smart City*, Springer Water,
https://doi.org/10.1007/978-3-030-94643-2_4

4.2 Urban Stormwater

Urban stormwater runoff is generated from urban housing, commercial and industrialized areas, paved areas, including open car parking lots, roads, highways, and bridges. During storm events, the inadequate capability of a surface for ponding and infiltration produces runoff. Once a land-use change occurs towards urban city areas, significant alterations result in the hydrology cycle. The urbanized area directly converts into the surface runoff in a storm event compared to the unpaved surface or the forest, as described in Chap. 2. Conventionally, stormwater in the city area is managed through a massive curb-gutter, catch basin, and storm drainage to fast transfer to the receiving watercourses. Typically, sewer systems convey stormwater runoff by either separate storm sewers or combined sewers.

Separate storm sewer systems dispose only of stormwater runoff to receiving watercourses without treatment. In the developed cities, sanitary sewer flows transfer to a particular sewer system towards public/metropolitan wastewater treatment plants.

A *combined sewer system* receives massive stormwater runoff in addition to sanitary sewer flows for disposal. Municipal wastewater treatment plants treat flows from combined sewers before disposing to receiver watercourses. However, in heavy storm events, increased storm runoff often exceeds the wastewater storage and treatment capacity. On the other hand, *combined sewer overflows (CSOs)* are the untreated stormwater and sanitary wastewater directed to receiving watercourses. *Urban runoff* is the surface runoff generated from storm events due to urbanization. This runoff is a significant source of floods and polluted water in the urbanized area, and these are listed in Table 4.1. The following constituents are usually investigated in urban runoff:

- Total Suspended Solids (TSS)
- Biochemical Oxygen Demand (BOD)
- Chemical Oxygen Demand (COD)
- Total Phosphorus (TP)
- Soluble Phosphorus (SP)
- Total Kjeldahl Nitrogen (TKN)
- Nitrate and Nitrite (N)
- Total Copper (Cu)
- Total Lead (Pb)
- Total Zinc (Zn)
- Coliform bacteria

The *Nationwide Urban Runoff Program (NURP)* was conducted between 1978 and 1983 by US EPA. Based on several storm events, land-use variability sites, geographical locations, NURP has studied urban runoff characteristics and their adverse effects on water quality. Thus, the management and practices to handle pollution loads from urban runoff are illustrated in Table 4.2.

Table 4.1 Sources of contaminants in stormwater runoff in the urban environment (Oberts et al. 1989; USEPA 1999; Burton and Pitt 2002)

Contaminants	Sources of contaminants
Sediment and floatable (i.e. TSS)	Roads, driveways, pedestrian way, road maintenance, car washing, corrosion of vehicles, construction materials, during the dry season, atmospheric deposition of organic matter from plants and animals
Pesticides and herbicides	Lawns and gardens in a residential area, roadsides, utility right-of-ways or public areas, commercial and industrial areas
Organic Materials	Lawns, gardens, improper handling of municipal solid waste
Metals (Cd, Cu, Zn, and Pb are commonly reported; also Co, Cr, Fe, Mn, Ni, and Platinum group elements often noticed in urban runoff)	Automobiles (tire wear, road wear, lubricants, auto body and engine corrosion, brake linings), rusts of *road furniture* or fixtures on the road surface, construction materials, atmospheric deposition, combustion processes, bridges, industrial areas, soil erosion
Oil and grease/hydrocarbons	Roads, driveways, parking lots, vehicle maintenance areas, and gas stations
Pathogens (virus, bacteria, fungi, and parasites)	Lawns, roads, leaky sanitary sewer lines, sanitary sewer cross-connections, animal waste, septic systems
Nitrogen and Phosphorus	Lawn fertilizers and wastes, atmospheric deposition, automobile exhaust, soil erosion, detergents, degradation of organic matters, animal and human waste, CSOs

Table 4.2 Median event-based mean pollutant concentrations for urban areas (USEPA 1983)

Pollutant	units	Residential		Mixed		Commercial		Open/Non-urban	
		Median	COV	Median	COV	Median	COV	Median	COV
BOD	mg/l	10	0.41	7.8	0.52	9.3	0.31	–	–
COD	mg/l	73	0.55	65	0.58	57	0.39	40	0.78
TSS	mg/l	101	0.96	67	1.14	69	0.85	70	2.92
Total Lead	μg/l	144	0.75	114	1.35	104	0.68	30	1.52
Total Copper	μg/l	33	0.99	27	1.32	29	0.81	–	–
Total Zinc	μg/l	135	0.84	154	0.78	226	1.07	195	0.66
Total Kjeldal Nitrogen	μg/l	1900	0.73	1288	0.50	1179	0.43	965	1.00
Nitrate + Nitrite	μg/l	736	0.83	558	0.67	572	0.48	543	0.91
Total Phosphrous	μg/l	383	0.69	263	0.75	201	0.67	121	1.66
Soluble Phosphorus	μg/l	143	0.46	56	0.75	80	0.71	26	2.11

COV = Coefficient of variation

4.3 Stormwater Management

For stormwater management, the concept of *Low Impact Development* (LID) and *Green Infrastructure* or *Green Stormwater Infrastructure* (GSI) is mainly used in the US. Similarly, *Sustainable urban Drainage Systems* (SuDS) are the most desirable option in the UK, decentralized stormwater management in Germany, *Low Impact Urban Design and Development* (LIUDD) in New Zealand, and *Water Sensitive Urban Design* (WSUD) in Australia (Hoyer et al. 2011).

Low Impact Development (LID) is a stormwater management approach to reproduce the natural hydrological processes through integrated planning, designing, and practices. Thus, distributed stormwater management is ensured through the training of on-site natural features. Conventional stormwater management systems consist of moving stormwater off-site through curbs, pipes, ditches, and ponds. LID approach mimics the pre-disturbance hydrologic processes of infiltration, recharge, evaporation, and transpiration of the concerned site with the changing landscape (as shown in Fig. 4.1). Thus, stormwater manages on-site in the LID approach, the rate and volume of predevelopment stormwater remain unchanged to the receiving waters.

Green Stormwater Infrastructure (GSI) or *Green Infrastructure* is a sustainable approach based on wet weather management. Therefore, to maintain or restore natural hydrology, this includes infiltration, evapotranspiration, capture, and stormwater reuse, as shown in Fig. 4.2. Green Infrastructure consists of the best possible green space in urban planning and benefits from these green spaces (Fletcher et al. 2015). Thus, this approach covers the total urban water cycle and supports the rainfall-runoff management principles.

Sustainable Urban Drainage Systems (SuDS) contains the technical knowledge and tools applicable to sustainable stormwater disposal than conventional practices. SuDS aim to replicate the natural conditions at a proposed urban site by conserving surface water, reducing inflow rates to the receiving waters, and improving water quality. In SuDS, there are the following steps to ensure water quantity, water quality, biodiversity, and amenity:

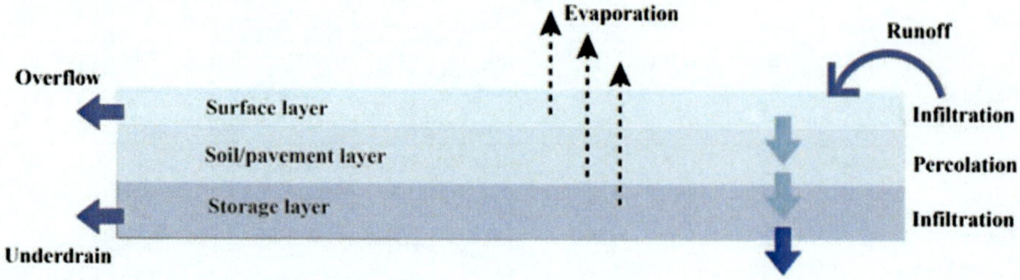

Fig. 4.1 Conceptual diagram of a LID process (James 2012)

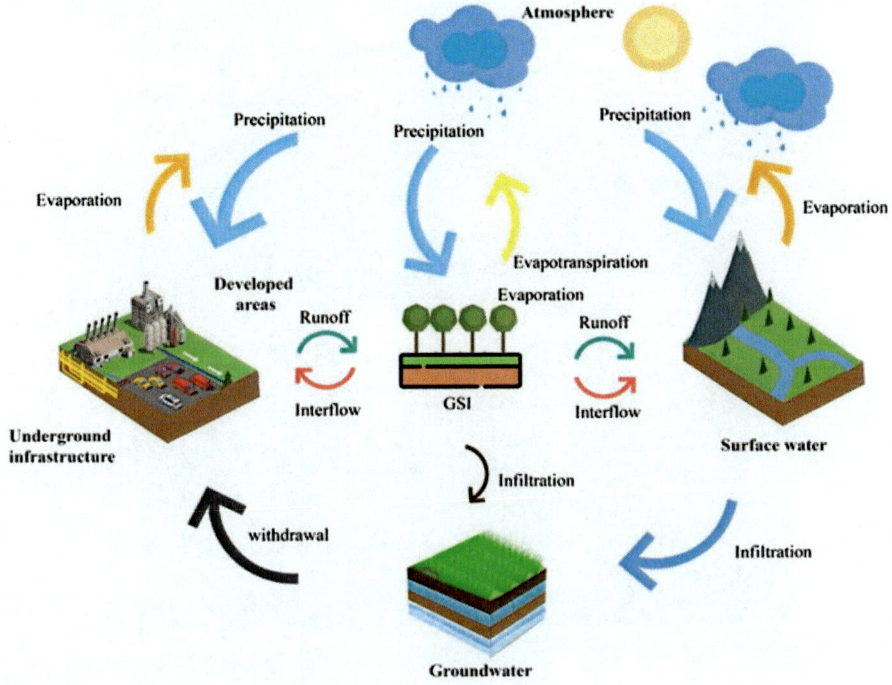

Fig. 4.2 Conceptual diagram of a GSI

Source control decreases inflow through the drainage/stream network by capturing runoff on catchment for harvesting or reuse as irrigation and consequent evapotranspiration viz. green roofs.
Pre-treatment removes pollutants from surface water before disposal using vegetated swales, ditches, or filter trenches.
Retention systems slow down the required time to transfer the surface runoff to watercourses through storages, i.e., ponds, retention basins, wetlands, etc.
Infiltration systems can mimic natural recharge through enhanced soil moisture absorption by providing infiltration trenches and soakaways.

Details on green roofs, bioretention swales, and basins, infiltration trenches describe in this chapter. A conceptual diagram of SuDS has presented in Fig. 4.3.

Water Sensitive Urban Design (WSUD) is a practice to achieve water balance by conserving water quantity and quality and maintaining environmental water requirements. The WSUD approach treats stormwater before entering a waterway or considering reusing it for another purpose. Thus, this approach conserves water quantity, supply, and quality and protects and preserves amenities and functions (Fig. 4.4). WSUD considered the following techniques:

- Using water-efficient appliances and landscapes reduces potable water consumption;

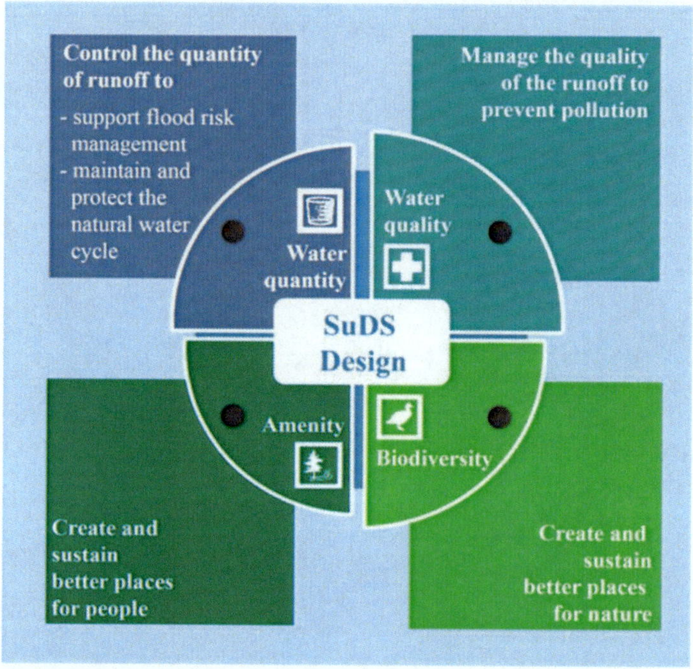

Fig. 4.3 Conceptual diagram of SuDS

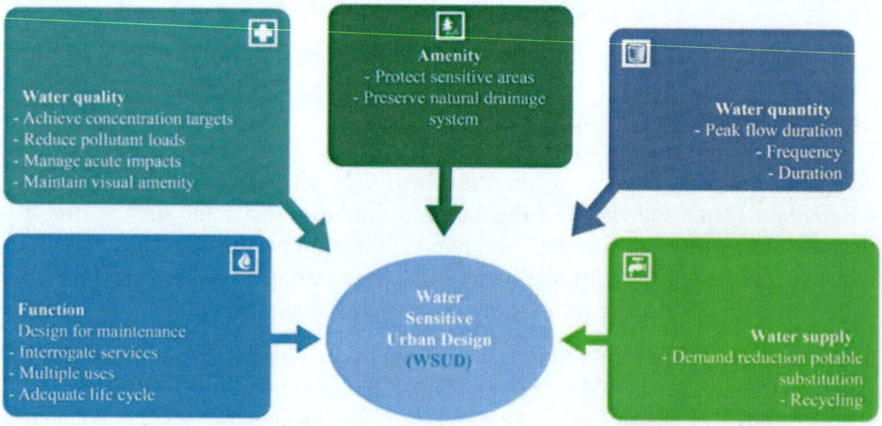

Fig. 4.4 Conceptual diagram of WSUD

- Conserve potable supplies, greywater reuse, as well as localized wastewater treatment practices;
- Reuse, storage, and infiltrate urban runoff rather than drainage system augmentation to allocate environmental water requirements;
- To protect the water-related ecosystem, social and heritage prices by minimizing the ecological footprint of a proposed project comprising integrated water resources management; and
- To withstand the adverse effects of climate changes through flexible institutional arrangements, long-term planning, and a wide range of water sources are handled by centralized and decentralized water infrastructure.

Low Impact Urban Design and Development (LIUDD) emphasize environmental vulnerability due to city developments and ensure sustainability. This approach enhances sustainable cities through integrating human activities and natural processes. Thus, the local climate and water regulation conserved recreation, amenity, and mental and health well-being. LIUDD refers to planning, techniques, and implements to ensure human urban development activities utilize rather than damage or destroy natural processes.

Best Management Practice (BMP) comprises techniques implemented to prevent pollution and urban runoff due to LULC. BMPs are designed to reduce the volume of stormwater, peak flows, and nonpoint source pollution. The working principle is to ensure natural hydrological processes and safe disposal based on biological and chemical actions. For stormwater management, there are two approaches, i.e., structural and non-structural. *Structural devices* are constructed using silt barriers, rock filter dams, fiber rolls, and sediment traps. Thus, typical structural devices are extended detention ponds, wet pond/detention ponds, infiltration basins, porous pavement, and water quality inlets. *Non-structural* approaches include reformed landscaping, scheduling land cover alteration, or street sweeping. Stormwater management BMPs categorized into four basic types (Alberta Environmental Protection 1994; USEPA 2016), and these are:

Storage practices comprise ponds, reclamation, and design for green infrastructure.

Vegetative practices consider buffers, networks, green roofs, stormwater wetlands design, and engineering.

Filtration/Infiltration practices include filtration, rain gardens, porous pavement, civic infrastructure and design, functional stormwater management.

Water-sensitive development covers improved location, open space design, and LID.

4.4 Developments in Stormwater Management

Since ancient times, water management was prioritized to meet basic human needs of clean drinking water supply, sanitation, hygiene, protection against water-related disasters, and progressively embraced conventional stormwater management. The traditional approach refers to a massive and expensive centralized infrastructure system. In approximately 600 BC, a stormwater drainage system was built in Cloaca maxima in Rome (Fardin et al. 2014). In the 19th century, due to the industrial revolution and rapid urbanization, underground stormwater drainage was adopted in Europe and the US (Burian and Edwards 2002). Meanwhile, *combined sewers* have been practiced in many cities to drain in nearby receiving water bodies. Due to financial and environmental benefits, integrated sewer approaches were preferred to minimize flood and urban pollution. In the 1920s, stormwater was considered wastewater, not a resource, and should be disposed of outside the cities (Durrans 2003). Thus, disposing of stormwater to the adjacent water bodies with gravity inspired developing urban drainage systems and, therefore, floods mitigation (Saraswat et al. 2016). In this connection, initiations were observed to develop design guidelines for urban drainage systems to manage *minor floods* (i.e., 2–25 year storm events) to *major floods* (i.e., 100-year storm events) (Grigg 2012). In the early 1980s around the world, conventional stormwater handling seemed inadequate to secure urban setup to eliminate urban floods; emphasis was instead given to utilizing rainwater along with different management and technical skills (Fig. 4.5). The "Thai Jar Programme" was a government program in the 1980s and a turning point for developing and promoting rainwater harvesting. Therefore gradual actions are recorded worldwide with the stormwater management policy, formal institutional arrangements, regulatory agencies, legislation developments of design manual/code of practices, and many cases even own computer modeling tools (Fig. 4.5). Thus, this advancement progressed from the conveyance-oriented concept of stormwater management to source control-oriented, for instance, the relevant event titled **A**ctive **B**eautiful **C**lean (ABC) water program in Singapore.

Thus, according to the global history, design, and implementation, stormwater management primarily focuses on the three-folds specificity following Fletcher et al. (2015) as shown in Fig. 4.6. The specificity extends from specific techniques, then conceptual and broad principles. Thus, the primary focuses are these techniques include only urban stormwater management or the whole urban water management.

4.5 Stormwater Management Technics

Stormwater management plays a critical role in urban life; countrywide, the practice might be any options, i.e., LID, GSI, SuDS, WSUD, LIUDD, or BMPs. There are similarities in the adopted techniques; the standard methods are described in this section.

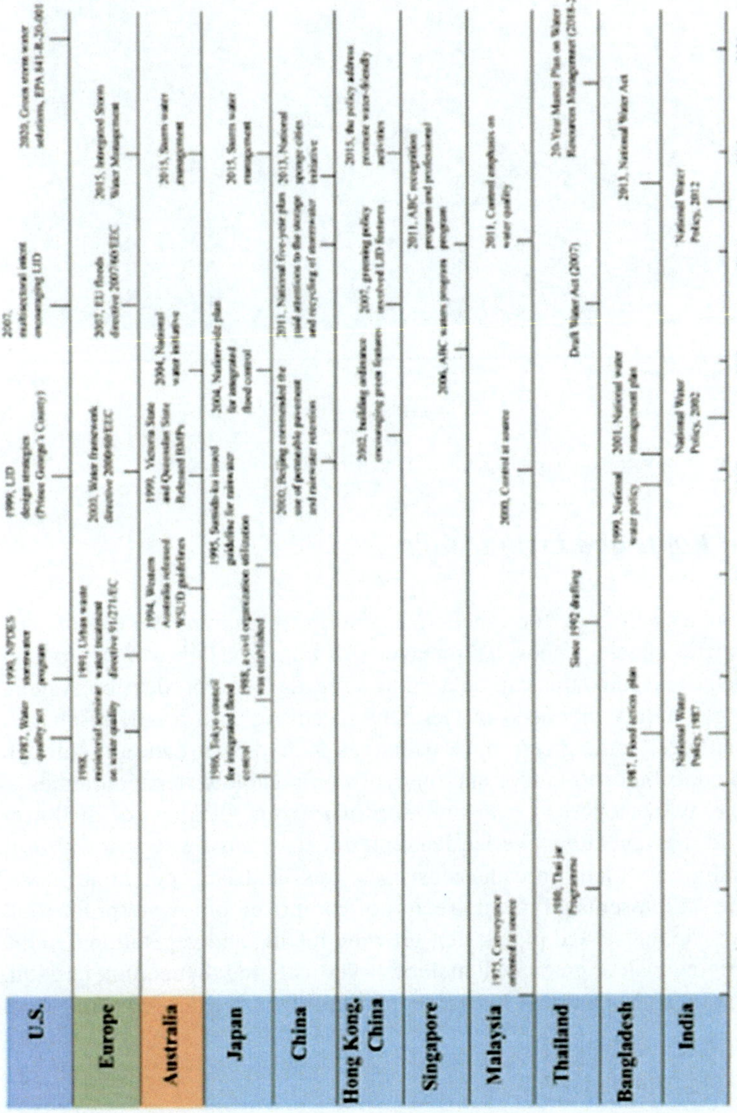

Here, NPDES = National Pollutant Discharge Elimination System

Fig. 4.5 Advancement in stormwater management around the world after Chang et al. (2018)

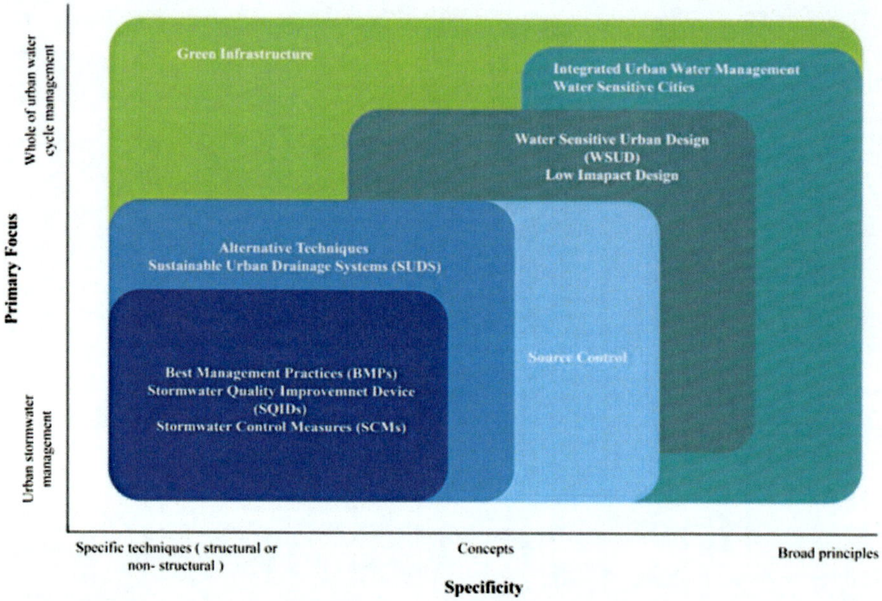

Fig. 4.6 Classification of stormwater management (Fletcher et al. 2015)

4.5.1 Green Roofs and Living Walls

A green roof involves a high-performance waterproofing membrane of the roof deck and plant root barrier system, disposal, filter layer, substrate, or a lightweight growing medium or vegetation. Thus, this can be used as a rooftop food production system. This system is commonly practiced in Germany to ensure zero runoff discharge. Here, rainfall captures on the green roofs, recharges groundwater naturally through infiltration, and contributes to indoor and outdoor non-potable water requirements. Green roofs are of two categories: *extensive roofs* of 150 mm thickness or shallower designed to avoid the vegetative overburden; and thicker *intensive vegetative roofs* merges into suitable on-structure arcade landscapes, grassland, large perennial plants, and trees. Figure 4.7 presents a typical green roof comprised of a waterproof roof, storage, drainage via the slotted pipe, filter, growing media, and vegetation. On the other hand, the living wall or green wall includes a vertical growth medium, i.e., soil, substitute substrate, and integrated hydration and fertilizer delivery system (Medl et al. 2017).

The advantages of green roofs are:

- Reduction of stormwater runoff: green roofs can absorb up to 80% of the rain
- Water quality improvement
- Alleviation of urban heat-island effects
- Offers a prolonged service life of roofing materials

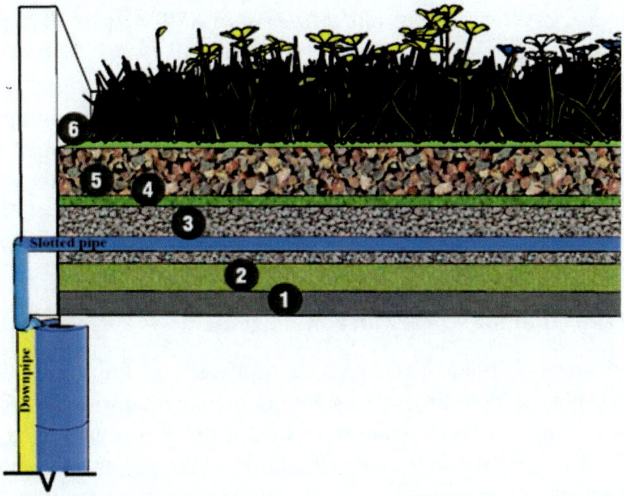

1: Insulation and waterproofing roof deck
2: Protection and storage layer
3: Drainage and capillarity layer
4: Root permeable filter layer
5: Extensive growing media
6: Vegetation

Fig. 4.7 Details on a green roof system

- Energy conservation
- Minimize echo and acoustic transmission; and
- Modification of the environmental aesthetics in both office and residence.

Green roofs often admit solar panels and perform well in association with LID. Similarly, a broad range of benefits in terms of environmental and socio-economical are achieved, and these are:

- The plants in a living wall purify the air;
- Plants absorb 50% sunlight, and 30% reflects, thus, reduces the ambient temperature;
- This roof reduces ambient noise of the building; therefore, a reduction of 8 dB could be achieved; and
- The peak load on the sewage system minimizes the rainwater buffer, and so reduces flood risks.

(A) Relevant codes and standards

The capacity of green roofs for pollution reduction through retaining storm runoff and slow down runoff disposal has been addressed in relevant European codes; also, interest increased in the US codes and standards.

The International Code Council (ICC) code practices in the US for a design load of the green roof calculations, the 'wet weight' treats as the surplus dead load. Additionally, requirements for the live load to maintain foot traffic and ensures pedestrian access. ICC also describes the criteria for the parapet walls.

American Society for Testing and Materials (ASTM) provides guidelines and testing processes for green roof construction materials.

Guideline for the Planning, Execution and Upkeep of Green-Roof Sites (Richlinien für die Planung, Ausführung and Plege von Dachbegrünung) in Germany by German FLL (Forschungsgesellschaft Landschaftsentwicklung Landschaftsbau e.V.) is the widespread building strategies for green roof. This strategy guides roofing material testing standards for almost 90% of the climate zones in the US. German FLL influences the other available standards and guidelines for green roofs worldwide, i.e., FM Global, ASTM, NRCA, SPRI, etc.

(B) Plant selection for stormwater management

Green roof designers should maximize the purposes of the green roof/green wall while selecting plants, including stormwater management through storm runoff quantity reduction, elimination of contaminants from the stormwater runoff, aesthetical consideration, drought tolerance, and availability of the plants. Higher water requirement plant species act as an interface through uptake water from the substrate and return into the atmosphere. Thus, higher water loss ensures significant water movement and increases naturally cooling surroundings. Typically the plant selections are:

- During the storm events to soak up water and contaminants removal, the selected plants should accumulate nutrients from available water—for instance, Herbaceous or shrubby species.
- For an impressive aesthetical view, selected plant species could be all-season fanciful foliage plants or/and flowers. For example, evergreen South Africa and southern Africa originated Leonotis leonurus (i.e., lion's tail and wild dagga), east Asian native Agastache rugose, or Korean mint.
- To survive during droughts, plants from shallow soils and rock outcrops usually survive in extended dry periods and consume excess rainfall in storm events. These are Dianella revoluta, Stypandra glauca, and Arthropodium milleflorum. The selection of drought-tolerant species planting in layers is another approach. Bulbine bulbosa, Senecio spathulatus, and other seasonally dormant species could be planted along with perennial species.

(C) Typical reasons for green roof failure

- Design and implementation deficiencies, inexperienced green roof professionals, or misunderstand and ignorance could miss incorporating the essential building and site requirements;
- Penetration flashing problems within the roof membrane;
- Failures happen due to faulty and improper flashing construction;
- The number of downspouts, inundation within parapet walls, and roof slope are essential issues to avoid ponding in storm events;
- Weathering has adverse action to deteriorate roof;
- Wind and airborne debris affect green roofs. Hurricane or tornado or cyclone induces wind collapse a green roof; and

- Care should be taken during placing the green roof/living wall attached components, including Heating, Ventilation, and Air Conditioning (HVAC), solar panels, antennas, flag poles, bracings, etc.

4.5.2 Rain Gardens

A rain garden or bioretention cell is a sink with porous, free-draining soil and planted with vegetation to survive against temporary flooding. Rain gardens are designed to reduce the volume and contaminants through replicating the natural water retention of undeveloped land (Fig. 4.8a) and rain garden planter or *planter box* for paved areas (Fig. 4.8a). For every 50 m^2 area of runoff, the suggested rain garden should be 1 m (Melbourne water 2009). The design considerations are:

For undeveloped land (Fig. 4.8a).

- A rain garden maintains allowance from the adjacent and neighbouring construction, i.e., both of the offsets A and B should be a minimum of 3 m.
- The garden should comprise of splash rock to slow down the inflow as well as reduce erosions.
- A berm should be provided around the garden.

For paved areas

The rain garden planter comprises bioretention soil, gravel bed, and existing subsoil. Inflow from impermeable surfaces allows the plant and drain to be absorbed. Then, rest flows over the bioretention soil included filter medium layer (loamy sand) and transition layer (sand), and passed over a gravel drainage layer with submerged zone is contains a slotted pipe–grated overflow pit. This arrangement is often placed on the existing subsoil, as shown in Fig. 4.8b, following practices in Australia, the UK, and other parts of the world (Melbourne water 2009; Bob et al. 2012). Otherwise, the base and sides of rain gardens are usually lined to the top of the overflow pipe. Thus, permanent retention volume provides water storage to sustain plants between rainfall events and protect nearby building footings. Typical design considerations are:

- Planting should be as per local guidelines by assessing the site and situation, specific conditions (plant groups with the same pH and soil moisture requirements), and aesthetic possibilities of enlivening the mixed plants.
- If the berm is designed to hold back water during a storm, this should be approximately 300 mm × 100 mm, and well compacted.
- At the center, the excavation base is 150–450 mm below the ground level.
- A channel needs to dispose of excess water in the drain direction. The channel needs to be permeably required with a 150 mm wide slot and gravel filling.
- Planter bottom of 50 mm is designed with stones and gravel, providing a 'fleece' over the gravel top, and the rest is filled with water-absorbent and free-draining

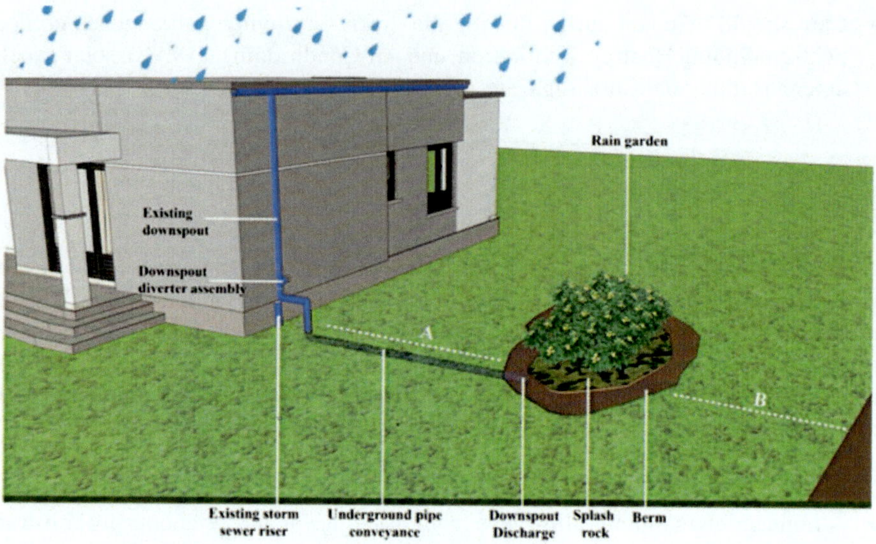

(a) Rain garden in association to the rainwater harvesting

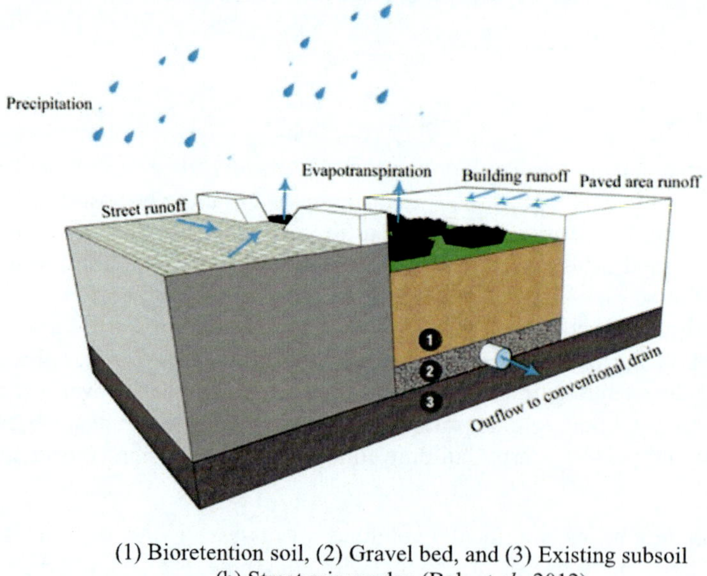

(1) Bioretention soil, (2) Gravel bed, and (3) Existing subsoil
(b) Street rain garden (Bob *et al.*, 2012)

Fig. 4.8 Details on rain garden

soil. This construction should avoid clay but emphasize the inclusion of more organic material.

Typical reasons for rain garden failure.

Design, implementation, and maintenances are essential to achieve the benefits of a rain garden. However, often following problems are experienced:

- *Lack of underdrain*: although rain gardens are designed to infiltrate storm water into the soil, excess runoff could fill up faster than disposes to drain.
- *Clogging the perforations*: clay and silt could block the underdrain pipe. To fix this, flush the underdrain pipe with a hose through the outlet structure.
- *Sediment washes into the rain garden from surrounding surfaces*: filling in all the air pockets in the mulch prevents infiltration. Protection through silt fence or filter stock till the neighbouring areas are stabilized using vegetation or paving.
- *Water seepage issues*: heavy construction vehicles accidentally compress subgrade soil to the degree that restricts water from infiltrating.

4.5.3 Bioretention Swales

Bioretention swales are designed for stormwater runoff conveyance and storage in a shallow, vegetated, landscaped depression with a side slope. Surface runoff undergoes water quality treatment while passing through the bio-retention area. These swales are usually implemented along the local and collector roads within residential areas, shared paths, medians, roundabouts, or other unused right-of-way areas. Swales require low construction costs but use more space than planters to handle low to moderate urban runoff. Typically, three layers exist in a bioretention swale, i.e., mulch or gravel layer, filtration layer, and drainage layer (Fig. 4.9). Following the design process should be taken for a bioretention swale:

- Landscape design needs to address stormwater quality objectives, thus, also confirm treatment performance;
- Predicting design flows for the swale component and treatment flow delivers urban runoff to a bioretention filter. Then the non-scour discharge is uniformly distributed over the entire filter media surface;
- Dimensioning the swale component, verifying size and configuration for treatment;
- Design inflow systems, i.e., size overflow pit (field inlet pits);
- Verify design:
  - Above ground components:
    - Velocities verification
    - Design of inlet zone and overflow pits
    - Verify design flow

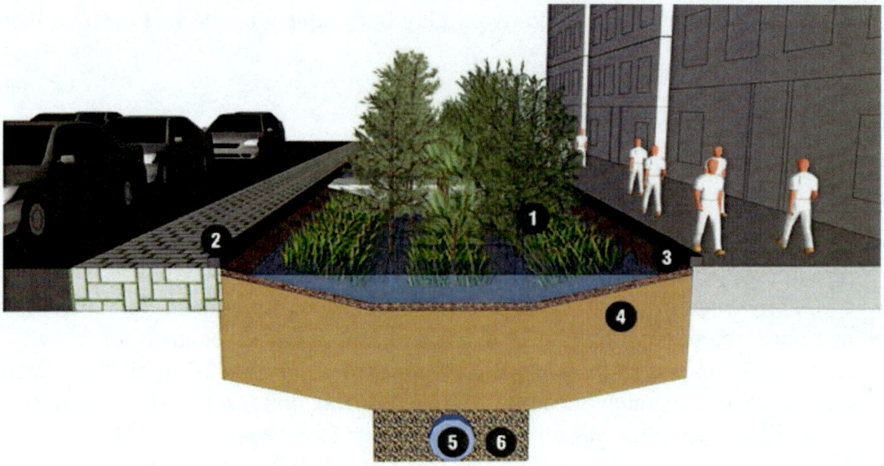

1. Top of extended detention (TED) or maximum ponding depth
2. Kerb
3. Mulch or gravel layer
4. Filtration layer
5. Perforated pipe
6. Drainage layer between 4 and 6, geotextile fabric

Fig. 4.9 Bioretention swale in the city

- Below ground components:
  - Ensure prescribed soil media layers, i.e., filter, transition, and drainage
  - Design and verify underdrain capacity
  - Verify bioretention lining requirements
- Allowances to exclude traffic load on swales.
- Plant species and their planting densities should be adopted from the available standard or codes.
- Provision for maintenance:
  - Maintenance is the primary reason for premature.

Stepwise bioretention swale design is as follows:

Step 1:*Conceptualize the desired treatment performance* along with the infiltration of excess runoff. Typically, treatment considerations include the conventional wastewater pollution parameters, i.e., TSS, TN, and TP.

Step 2: *Design flow determination* is based on both minor and major floods. For a relatively small catchment area, usually, rational method design procedures are considered. Thus, time of concentration (Tc), rainfall intensity (I), and design runoff coefficient (C) should be required for computing peak design flows.

Step 3: *Dimensioning swale components* include determining the width and side slopes, the maximum length, and capacity. The *maximum swale width* is required for the swale design. Site constraints affect *side slopes for swales* in parks, open spaces, or median strips, and the slopes range between 1:10 to 1:6. The *maximum length of a swale,* i.e., the distance along a swale before an overflow pit, is essential to dispose of through an underlying drain. Applying Manning's equation following Eq. 2.20a stated in Chap. 2 for the swale flow capacity:

$$Q = \frac{A.R^{2/3}.S^{1/2}}{n} \tag{4.1}$$

If the flow depth in a swale is lower than the vegetation height (preferable for treatment), recommended 'n' is between 0.15 and 0.4, along with the values from Table B.2, Appendix B. On the contrary, if the flow depth is double or more, the vegetation height recommended 'n' is 0.03.

Step 4: *Designing inflow systems* for the bioretention swale utilizing both point and nonpoint sources of urban runoff, i.e., pipe outfalls and from flush curbs on a road, respectively. Additionally, there are also combinations of these inflow pathways.

Step 5: *Designing bioretention components* is based on a minimum of two types of soil media for the bioretention swale, i.e., filtration layer (0.4–1 m sandy loam) and drainage layer (0.15 m coarse sand/gravel). The transition layer should be of geotextile fabric to restrict filter media's washout. The recommended thickness of the transition layer is 150 mm.

The maximum spacing (center to center) of the perforated pipes is 1.5 m; thus, the distance water travels horizontally through the drainage layer of the filtration media. The maximum filtration rate through the bioretention filter computed by Darcy's equation after Eq. 2.31 (described in Chap. 2) is as follows:

$$Q_{max} = K_{sat}.L.W_{base}\frac{h_{max} + d_{filter}}{d_{filter}} \tag{4.2}$$

where

Q_{max} = Maximum filtration rate (m^3/s)

K_{sat} = Saturated hydraulic conductivity for the filter layer (m/s); values could be extracted from Table B.1, Appendix B

W_{base} = The base width of the ponded cross-section above the soil filter (m)

L = Length of the bioretention zone (m)

h_{max} = Ponding depth above the soil filter (m)

d_{filter} = Depth of filter media (m)

The under-drainage via perforations system should be sufficiently capable of collecting and conveying the maximum infiltration rate. Here, half of the perforation is allowed to be blocked. The slotted pipes should undergo several checks to ensure adequate size:

- Adequate numbers of slots are within the pipe to convey the maximum filtration rate.

- The flow capacity of the perforations is calculated using the basic orifice formula, i.e.:

$$Q_{perf} = B.C_D.A_o\sqrt{2gh} \tag{4.3}$$

where

Q_{perf} = Flow-through perforations (m^3/s)
C_D = Orifice discharge coefficient, 0.6 to 0.65 (Medaugh and Johnson 1940)
A_o = The total area of the orifice (m^2)
h = Maximum depth of water above the pipe (m)
B = Blockage factor (suggested for 50% blockage of the perforations)

- Through the transition layer, the materials in the filter layer could be restricted washed away into the perforated pipes

Step 6: *Verifying design* through potential scour velocities and depths should be checked. Potential scour velocities of bioretention swale should be less than 0.5 m/s and 2.0 m/s for minor flood and major flood discharge, respectively (DPI, IMEA and BCC 1992). Bioretention swales should satisfy the following general safety criteria to ensure public accessibility at the intersections, footpaths, and bicycle pathways:

$$depth \times velocity < 0.4\ m^2/s$$

$$\text{TED} = 0.3\text{ m}$$

These checks are usually practiced in Australia (DPI, IMEA and BCC 1992). However, regional guidance should be applied if available for design verification. Once the velocity and depth checks are satisfactory, the next step is to confirm the treatment performance.

Step 7: *Sizing the overflow pit* requires flushing through the swale and bioretention system filter media while no extended detention is available. Also, the pit crest is raised above the filter media level for submerged or free-flowing conditions to launch the extended detention depth. The required weir's length could be determined using a broad crested weir equation for the free-flowing conditions. An orifice equation estimates the area between openings in the grate cover for submerged outlet conditions.

For free-flowing conditions (weir equation):

$$Q_{weir} = B.C_w.L.h^{3/2} \tag{4.4}$$

where

Q_{weir} = Flow into the pit (weir) under free overfall condition (m^3/s)
C_w = Weir coefficient, the typical value is 1.66

L = Weir Length (i.e., the perimeter of a pit) (m)
h = Flow depth above the weir or the pit (m)
For the drowned condition, the orifice equation (Eq. 4.3) should be used.

Step 8: Check with the traffic allowances on swales.
Step 9: Selection of plant species (few examples are provided in Appendix C).
Step 10: Provisions for maintenance.

Example problem 4.1 As shown in Fig. 4.10, an urban low-density residential area adopted a bioretention swale approach for stormwater management. The site for the bioretention swale comprises the arterial road and a service road divided by a 6 m wide median. Besides the service road, there is an adjoining low-density residential allotment, and the approximate depth is 30 m. Service road feeds the collector road. Overland flow slopes for both bioswales A and B are 1.3%. The soil type is clay. The site details are presented in Tables 4.3 and 4.4.

Design a bioretention swale *for this site.*

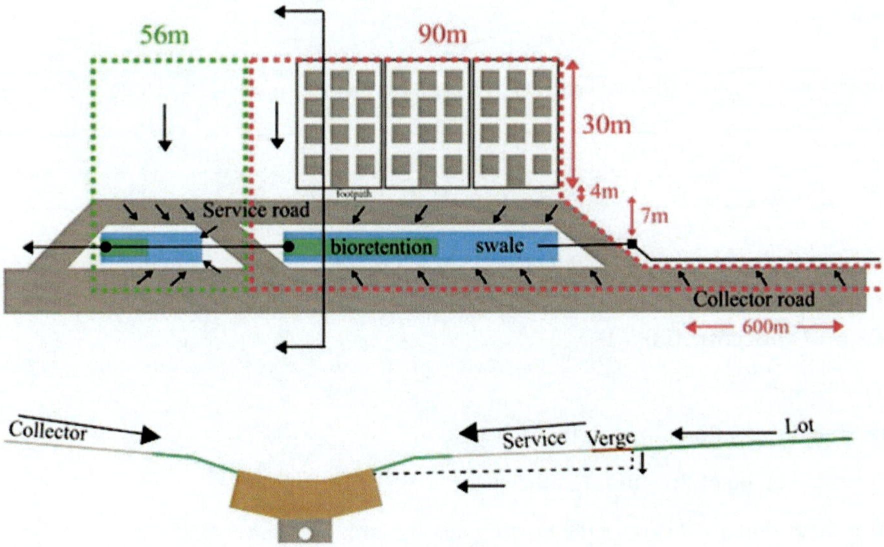

Fig. 4.10 Example problem 4.1 on bioretention swale

Table 4.3 Land use and land cover details

	Area	Collector road	Service road	Footpath	Swale
Bioswale A	90 m × 30 m	600 m × 7 m	90 m × 7 m	90 m × 4 m	90 m × 6 m
Bioswale B	56 m × 30 m	56 m × 7 m	56 m × 7 m	56 m × 4 m	45 m × 6 m
Percentage of imperviousness (%)	60	90	90	50	0

Table 4.4 Meteorological details

		10 years	100 years	Remarks
Average recurrence interval (ARI) rainfall (mm/hr)		100	150	Based on the regional IDF curve
Rainfall intensity (mm/hr) for different time of concentration, Tc	$T_c = 20$ min	120	200.5	
	$T_c = 15$ min	129.5	219	
	$T_c = 10$ min	140.5	239.5	
Runoff coefficient (from Table 2.6)	Bioswale A	0.75	0.94	Residential allotment
	Bioswale B	0.70	1.19	Asphaltic and concrete

Solution:

Step 1: *Conceptualize the desired treatment performance that* would meet the regional water quality standard. Based on the available allotment size, assume the bioretention cell area is one-fourth of the swale area. There is no thumb rule on this; this selection would be made based on the available land and facilities and the regional design guidance. Also, assume both of the cells would remove the TSS, TP, and TN.

Bioretention swale system	Bioretention swale area (m^2)	TSS removal (%)	TP removal (%)	TN removal (%)
A	135	95	80	50
B	$67.5 \approx 68$	95	80	50

Step 2: *Design flow determination* is based on both minor and major floods, i.e., for 10 year and 100 year ARI peak flows.

Time of concentration (T_C).

Bioswale A

Overland sheet flow through the site = 30 m

Overland channel flow through the swale = 90 m

The sheet flow path is over the lawn; assume Manning's 'n' = 0.25

(a) Minor flood

$T_c = t_{sheet\ flow} + t_{channel\ flow}$	
$= \frac{6.94(L.n)^{0.6}}{I^{0.4}S^{0.3}} + \frac{90}{0.3}$	Kinematic wave equation (Eq. 2.19) uses for sheet flow computation. Here, channel velocity has been taken as 0.3 m/s using Manning's formula
$= \frac{6.94(30\times0.25)^{0.6}}{(100)^{0.4}(0.013)^{0.3}} + \left(\frac{90}{0.3} \times \frac{1}{60}\right)$	
$= 13 + 5 = 18$ minutes	

(b) Major flood

$T_c = t_{sheet\ flow} + t_{channel\ flow}$	
$= \frac{6.94(30\times0.25)^{0.6}}{(150)^{0.4}(0.013)^{0.3}} + \left(\frac{90}{0.3} \times \frac{1}{60}\right)$	
$= 11 + 5 = 16$ minutes	

Bioswale B

Minor flood	Major flood
$T_c = \frac{6.94(30\times0.25)^{0.6}}{(100)^{0.4}(0.013)^{0.3}} + \left(\frac{56}{0.3} \times \frac{1}{60}\right)$	$T_c = \frac{6.94(30\times0.25)^{0.6}}{(150)^{0.4}(0.013)^{0.3}} + \left(\frac{56}{0.3} \times \frac{1}{60}\right)$
$= 13 + 3 = 16$ minutes	$= 11 + 3 = 14$ minutes

Design rainfall intensities (interpolating values stated in Table 4.4).

Bioretention swale	Minor flood		Major flood	
	T_c (min)	Rainfall intensity (mm/hr)	T_c (min)	Rainfall intensity (mm/hr)
Bioswale A	18	123.8	16	215.3
Bioswale B	16	127.6	14	223.1

Using the rational method (Eq. 2.18a), the obtained peak flows are:

	Bioswale A	Bioswale B
Minor flood	$Q = \frac{\left(0.75\times\frac{123.8}{1000}\times2700\right)}{360} = 0.70\ m^3/s$	$Q = \frac{\left(0.70\times\frac{127.6}{1000}\times1680\right)}{360} = 0.42\ m^3/s$
Major flood	$Q = \frac{\left(0.94\times\frac{215.3}{1000}\times2700\right)}{360} = 1.52\ m^3/s$	$Q = \frac{\left(1.19\times\frac{223.1}{1000}\times1680\right)}{360} = 1.24\ m^3/s$

Step 3: *Dimensioning swale components.*

(a) Swale width and side slope

The bioswale A and B components need to covey the required discharge viz.10 years and 100 years ARI flows. The designed swale dimensions for both bioswales are:

Swale base width = 1 m
Side slopes = 1V:5H
Max depth = 0.5 m
Moderate vegetation height = 300 mm
Manning's n = 0.04 (for flows above vegetation height); and

Longitudinal slope = 1.3%.

(b) Maximum swale length

Initially, the maximum length of *Bioswale A* is determined and then applied to the *Bioswale* B as this permits comparatively lower flow rates. The maximum length of *Bioswale A* requires calculating the total capacity of the swale by Manning's equation.

$$Q_{capacity} = 3.17\ m^3/s > 0.70\ m^3/s(Q_{10})\ \text{and}\ 1.52\ m^3/s(Q_{100})$$

Therefore, the swale capacity is adequate to convey all flows well over the Q_{100}, ensuring no flow inundates neighbouring road pavement. Thus, the maximum swale length for both bioswales is supposed to be longer than the 'actual' swale length.

Step 4: *Inflow Systems.*

Flow enters the bioretention swale systems of *'A'* and *'B'* through two mechanisms, i.e. (i) conveyed by perforated underground pipes from the collector road into bioswale A or runoff and (ii) direct runoff from the service road and footpaths. Allowance of flush curbs of 60 mm deposits sediment on the surfaces of the road. Scour protection for the pipe outlets should be using locally available materials to reduce local flow velocities and erosion.

Step 5: *Design Bioretention Component.*

(a) *Check the swale dimensions*

Steps 2 and 3 suggested dimensions are adequate to convey the design ARI flow. Then, the filter media selection (i.e., the saturated hydraulic conductivity) or extended detention depth are acceptable. Suppose the swale geometry cannot convey the minimum design floods on the neighbouring road pavements and fail to offer the minimum freeboard to the adjacent property. In that case, the swale geometry should be revised.

(b) *Selection of filter media*

The selection of the materials for filter and drainage layers should support the underdrainage system. For example, a slotted pipe of 1.5 mm slot width would be at a wash-out risk with the sand ($d_{50} \leq 1$ mm) selection as filter media. Therefore, to minimize this issue, multilayers could be applied to available standards or codes. Usually, multiple layers are comprised of sandy loam as the filter media (600 mm), coarse sand for the transition layer (150 mm), and fine gravel as the drainage layer (150 mm) (Fig. 4.10). Thus, the estimated saturated hydraulic conductivity of sandy loam filter media is about 180 mm/hr.

The median area is suitable for on-site stormwater treatment. The available space is more extensive to the site though in an elongated shape.

(c) *Drainage layer*

A 5 mm screening has provided in this layer.

(d) *Under drainage design and capacity checks*

The maximum infiltration rate over the filter media base and the head above the pipes are checked using Darcy's equation (Eq. 4.2):

$$Q_{\max} = 5 \times 10^{-5}.L.W.\frac{0.3 + 0.6}{0.6}$$

Here, maximum infiltration rate:

$$Bioswale\ A,\ Q_{\max} = 5 \times 10^{-5} \times 135 \times 1 \times \frac{0.3 + 0.6}{0.6} = 0.01\ m^3/s$$

$$Bioswale\ \ B,\ Q_{\max} = 5 \times 10^{-5} \times 68 \times 1 \times \frac{0.3 + 0.6}{0.6} = 0.0045 \approx 0.005\ m^3/s$$

Check for Perforations Inflow

Initially, the inlet capacity of the perforated drainage system should be free of clogs within the system. Usually, this assumes that 50% of the slots are blocked within a standard perforated pipe. Inflow rate is determined using the orifice equation (Eq. 4.3):

$$\begin{aligned}\text{Head above pipe (h)} = 0.95\ \text{m}[&0.6\ \text{m(filter depth)} \\ &+ 0.3\ \text{m(max. pond level)} + 0.05\text{(half of pipe diameter)}]\end{aligned}$$

Assume, sub-surface drains with half of all pipes blocked:

Clear opening = 2100 mm^2/m, hence blocked openings = 1050 mm^2/m (50%)
Slot width = 1.5 mm
Slot length = 7.5 mm
Number of rows = 6
Diameter = 100 mm

Number of slots per m = $\frac{1050}{1.5 \times 7.5} = 93.3$
Assume orifice flow conditions with the $C_d = 0.61$.
Inlet capacity per unit length of pipe:

$$\begin{aligned}Q_{perf} &= \left[0.61 \times (0.0015 \times 0.0075 \times 93.3)\sqrt{2 \times 9.81 \times 0.95}\right] \\ &= 0.0028\ \text{m}^3/\text{s}\end{aligned}$$

Considering pipe lengths for Cell A and Cell B are 65 m and 17 m, respectively. Thus, the inlet capacity is:

Cell A: $0.0028 \times 65 = 0.182\ m^3/s > 0.010\ m^3/s$ (maximum infiltration rate)

Cell B: $0.0028 \times 17 = 0.05\ m^3/s > 0.005\ m^3/s$ (maximum infiltration rate)

Therefore, a single pipe is capable of passing flows into the perforated pipe.

Perforated pipe capacity check

The flow rate in the perforated pipes is estimated using Manning's equation. Thus, the pipe capacity could be evaluated to convey the maximum filtration rate. Two 100 mm diameter perforated pipes are laid parallel at a longitudinal slope of 0.5% towards the overflow pit. Perforated pipe flow rate (for Manning's n of 0.02) is:

$$Q_{flow\ per\ pipe} = 0.0024\ m^3/s$$

Therefore, maximum infiltration rates for four perforated pipes for bioswale A and two perforated pipes for bioswale B are 0.01 m^3/s, and 0.005 m^3/s should be provided.

Drainage layer hydraulic conductivity check

The provision of flexible perforated pipes is used as a gravel filter pack; in this case, 5 mm gravel has been selected. This media is much coarser than the filter media, i.e., sandy loam. Therefore, a coarse sand-based transition layer of 150 mm thickness has been provided to minimize the washout risk.

(d) **Impervious liner requirement**

The available soils are clay to silty clays with a K_{sat} of 3.6 mm/hr in this site. The sandy loam media is suggested for the filter layer of 50–200 mm/hr K_{sat}. Therefore, the K_{sat} of the filter media is more significant than ten times the K_{sat} of the surrounding soils, and a waterproof liner is not required.

Step 6: *Design Verification.*

Applying Manning's equation, potential scour velocities need to be checked for the swale and the bioretention surface to ensure the following criteria:

- Minor flood: within 0.5 m/s
- Major flood: within 2.0 m/s.

Step 7: *Overflow Pit Design*

The overflow pits must dispose of minor floods safely from the bioretention systems and into an underground drainage system. Grated pits are provided at the downstream end of each bioswale. These pits are designed using a broad crested weir equation with the height above the maximum ponding depth and below the road surface, subtracting freeboard [i.e., 0.76 – (0.3 + 0.15) = 0.31 m].

The first trial used Eq. 4.4; here, the blockage factor is 0.5, the weir coefficient is 1.66, the required length of the weir is 2.6 m, and the flow depth above the weir is 0.31. Therefore, the equivalent pit area is 400 mm × 400 mm.

Check for drowned using Eq. 4.3, $C_d = 0.6$, and the $h = 0.31$ m.

Hence, the minimum pit area is 400 mm × 400 mm.

Step 8: Allowances to Preclude Traffic on Swales

Traffic controls use traffic bollards.

Step 9: Vegetation Specification

A mix of tufted grass and sedges is to be used (following Appendix C). For plantation, species of 300 mm height have been suggested. The landscape designer will finalize the actual species to be planted.

4.5.4 Bioretention Basin

The terms bio-filter and rain garden are also referred to as bioretention basins. Though rain gardens represent a small individual scale compared to bioretention basins of a more extended scale and can be used if there are several buildings and the lot is under single ownership (Fig. 4.11). The bioretention basin works combined into a car park and a local streetscape. Usually, a bioretention basin comprises filter media, transition layer, drainage layer, perforated pipe, and an overflow pit. Thus, bio-retention basins regulate water flow and facilitate water treatment.

The fundamental differences between bio-swales and bioretention basins are:

i. Basins aim to pond water compare to bio-swales, and
ii. Bio-swales are designed to transfer than the ponding strategy of bays.

Bioretention basin is based on the following design process:

Step 1: *Determine design flows* for both major and minor storm events.

Step 2: *Designing inflow systems* for scour protection, coarse sediment forebay, and street hydraulics.

Inlet scours protection ensures robust inflows from a point source (i.e., piped drainage) or a non-point source (i.e., roadside curb, open channel).

Rockwork is constructed to dissipate the energy of concentrated inflow, and the typical inlet scours protection is shown in Fig. 4.12.

Due to the absence of pre-treatments (through vegetated swale or buffer treatment), drainage basin stormwater runoff directly disposes to the bioretention basin. The required volume of forebay sediment storage is:

$$V_s = A.R.L_o.F_c \tag{4.5}$$

where

A = Drainage area (ha)

R = Capture or removal efficiency (viz. 80%)

L_o = Sediment loading rate ($m^3/ha/year$)

= 1.6 m^3/ha/year (for the developed urban area)

F_c = Anticipated cleaning frequency (year).

The forebay area, A_s is $\frac{V_s}{D}$. The forebay depth (D) is supposed to be within 0.3 m below the filter layer. The suitability for arresting 1 mm and greater particles is

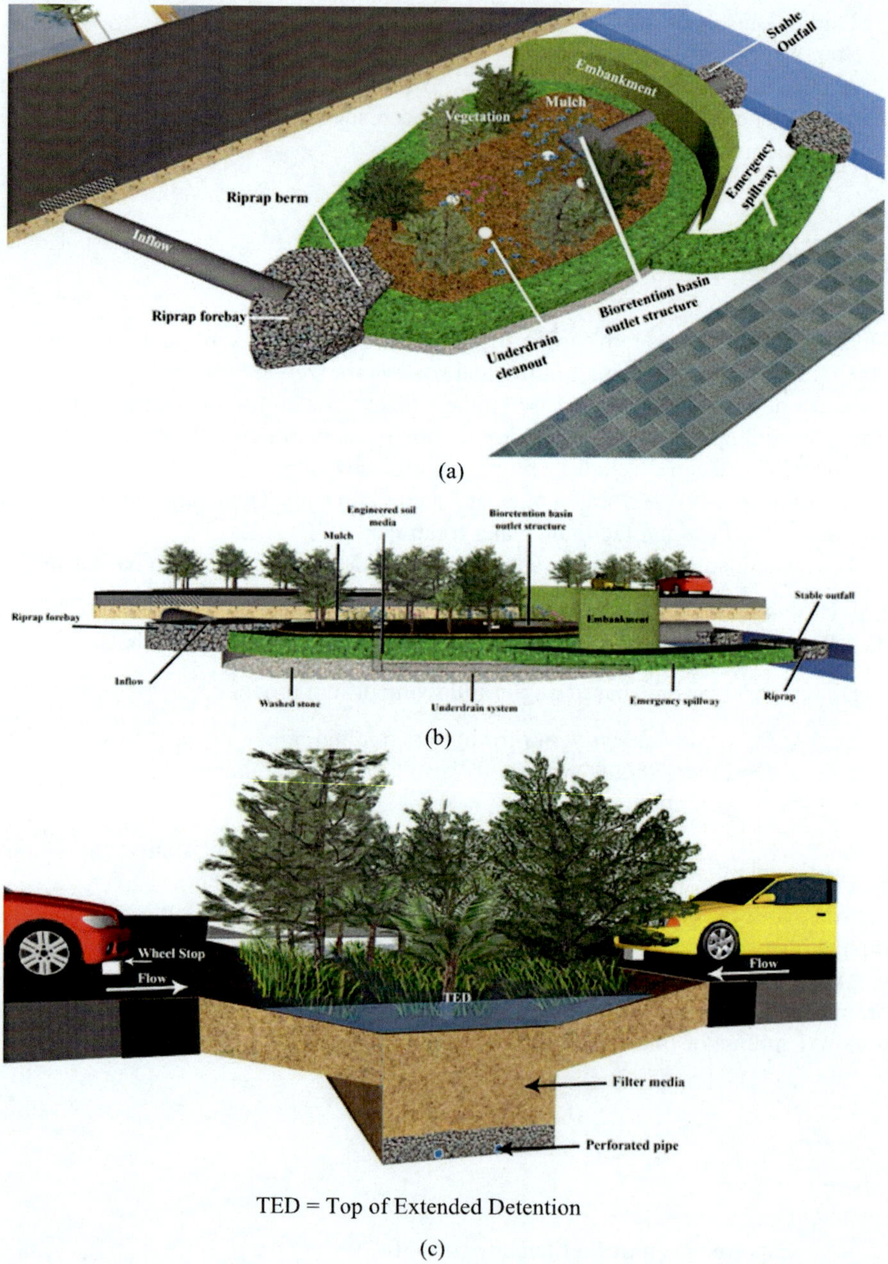

Fig. 4.11 **a** Bio retention basin plan, bio retention basin integrated into **b** car park and **c** local streetscape

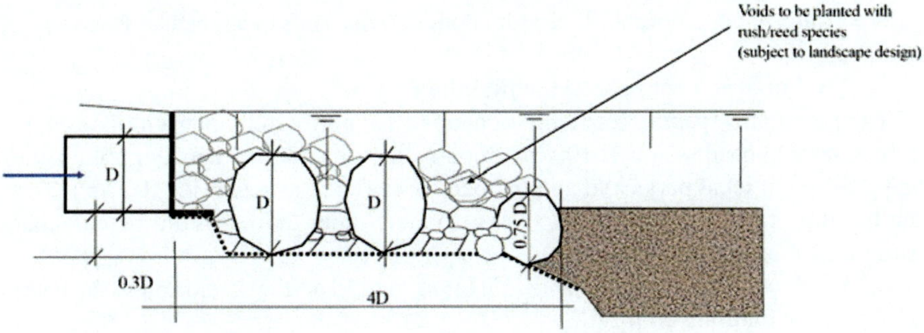

Fig. 4.12 Inlet scour protection detail for bioretention basin

evaluated using the following equation (Fair and Geyer 1954):

$$R = 1 - \left[1 + \frac{1}{n_{sc}} \cdot \frac{v_s}{Q/A}\right]^{-n_{sc}} \tag{4.6}$$

where

v_s = Sediment settling velocity (Assume 0.1 m/s for 1 mm particle)
$\frac{Q}{A}$ = Basin flow rate applied per surface area ($m^3/s/m^2$)
n_{sc} = Turbulence parameter (Assume, 0.5)

Manning's equation and a broad crested weir are applied to determine the curb opening width to allow flows through the bioretention basin.

Step 3: *Identify bioretention two to three filter layers, including* filter media, transition layer, and drainage layer. Regional guidance has followed to fix the layer depths; otherwise, the suggested depth would be 400–1000 mm. The minimum permissible depth for grasses and shrubs are 400 mm and 800 mm, respectively. Thus, minimize the complexities due to plant roots interfering with the slotted underdrain system. Then, the drainage layer is 150 mm to 200 mm thick. The recommended transition layer between these two is 150 mm thick.

Step 4: *Design under-drain and undertake capacity checks (optional) to ensure horizontal flow* through perforated pipes in the drainage layer to minimize drainage flow within the filter media.

The maximum filtration rate is the maximum flow rate through the paving system; Darcy's equation calculates this after Eq. 2.40 (described in Chap. 2) as follows:

$$Q_{\max} = KA\frac{h_{\max} + d_{filter}}{d_{filter}} \tag{4.7}$$

where

Q_{max} = Maximum filtration rate (m^3/s)
A = Area of bioretention basin (m^2)

K = Filter layer saturated hydraulic conductivity (m/s), values can be extracted from Table B.1, Appendix B

h_{max} = Pondage depth above the filter layer (m)

The maximum spacing (center to center) of the perforated under-drains is 1.5 m in bioretention basins (area < 100 m^2) for streetscape and low-density public zones; and 2.5–3 m in local parks and ample open space (i.e., basin areas > 100 m^2). Then, the filtration rate and capacity of the perforated under-drains would be calculated using Eqs. 4.2 and 4.3.

Step 5: *Requirement for impermeable lining* should be checked through the testing of K_{sat} of the natural soil profile.

Step 6: *Sizing of the overflow pit* depends on the requirement by the bioretention basin. The minimum permissible head is 100 mm above the pit crest to facilitate the design flow disposal into the overflow pit. Design consideration of the overflow pit is either submerged or free-flowing condition. Under the free-flowing condition, a broad crested weir equation determines the required weir length (i.e., Eq. 4.4). Submerged outlet condition uses an orifice equation to calculate the area needed between openings in the grate cover (i.e., Eq. 4.3).

Step 7: *The selection of the vegetation* should be based on regional guidance and consultation by landscape architects. Thus, the designed water treatment and the aesthetic view could be ensured.

Step 8: Verification checks with the scour velocities over the vegetation and the achieved water quality after passing the bioretention basin. Potential scour rates of bioretention swale should be less than 0.5 m/s and 2.0 m/s for minor floods and major flood discharge.

Typically, there are three stages involved for a bioretention basin construction and establishment, and these are stage 1: functional installation, stage 2: sediment and erosion control, and stage 3: operational establishment. *Stage 1* construction to provide temporary protection starts during landscape works and requires six months to 1 *year*. Then, *stage 2* ensures provisional erosion and sedimentation controller throughout the structure and requires one year to three years. In stage 3, removal of the temporary protections at the building phase following the design planting schedule, and overall, this stage takes 3 years to 3.5 years.

Example problem 4.2 A high-density urban residential area consists of the *road reserve*[1] and the adjacent three consecutive residential allotments (approximately 30 m in depth). The contributing site to each designed bioretention basins is the road reserve and the allotments. Runoff from adjacent properties disposes into the road gutter and, combined with road runoff, is transferred through a typical roadside gutter for approximately 30 m long towards the bioretention basin. The overland flow path length from the adjacent allotment to the gutter, is about 90 m. The road has a longitudinal grade of 1.3%, and the average slope of the neighboring allowances is 2%. The city authority would permit a 2 m width (i.e., perpendicular to the road

[1] A road reserve is a area contains the roads, footpaths, and associated features may be constructed for public travel.

Table 4.5 Land use and land cover information

	Allotments	Collector road	Local road	Footpath
Catchment area	90 m × 30 m	600 m × 7 m	90 m × 7 m	90 m × 4 m
Percentage of imperviousness (%)	60	90	90	50

Table 4.6 Rainfall – runoff details

		1 year	10 years	100 years	Remarks
Average recurrence interval (ARI) rainfall (mm/hr)		85.5	100	150	Based on the regional IDF curve
Rainfall intensity (mm/hr) for different time of concentration, T_c	$T_c = 20$ min	88	120	200.5	
	$T_c = 15$ min	90.6	129.5	219	
	$T_c = 10$ min	100.5	140.5	239.5	
	$T_c = 5$ min	120	155	265	
Runoff coefficient (from Table 2.6)		0.75	0.75	0.94	Residential allotment

alignment) for the bioretention basins. Design a series of bioretention basins to treat local catchment runoff using details from Tables 4.5 and 4.6.

The essential design elements of the bioretention basins are:

- Details on participating non-point sources to convey runoff into the basins;
- Information on the inlet is required to acquire erosion mitigation action;
- Configuration and design of an operating system are required to provide design allowance for minor flood (i.e., 10 years) mitigation on the local road;
- Details on the perforated under-drainage system;
- Specify available soil filter medium; and
- Landscape layout along with plant species.

Solution:

Step 1: *Concept design to confirm treatment performance* water quality treatment objectives would meet the regional (water quality) standard. Based on the available allotment size, three bioretention basins are planned with a center-to-center distance of 30 m, and each of them is 18 m^2. This selection would be made based on the available land and facilities and the regional design guidance as there is no thumb rule on this. To meet the objectives of treating stormwater on a sandy loam soil filter ($K_{sat} = 180$ mm/hr) of an extended detention depth of 300 mm. Also, assume the bioretention basins would maintain the reduction targets include 80% of TSS, 60% of TP, and 45% of TN.

Step 2: *Design flow determination* is based on minor (i.e., 10 years ARI peak flow) and major floods (i.e.100 years ARI peak flow).

Time of concentration (T_c): Contributing catchment area to each bioretention basin (i.e., within 30 m)

$$= \text{Allotments} + \text{Collector road/local road} + \text{Footpath}$$
$$= (30 \times 30) + (30 \times 7) + (30 \times 4)$$
$$= 1230 \text{ m}^2$$

Adjacent allotments contributed overland sheet flow, the flow path = 30 m.
Overland channel flow through curb and road gutter to each bioretention basin

$$= 30 \text{ m}$$

A sheet flow path over the lawn, assume, Manning's n = 0.25.
The slope of the adjacent allotments = 2%
Road longitudinal slope = 1.3%

a. Minor flood

$t_c = t_{sheet\ flow} + t_{channel\ flow}$	
$= \frac{6.94(L.n)^{0.6}}{I^{0.4}S^{0.3}} + \left(\frac{30}{0.3} \times \frac{1}{60}\right)$	for sheet flow computation apply of Kinematic wave equation (Eq. 2.21). Here, channel velocity has been taken as 0.3 m/s using the Manning's formula
$= \frac{6.94(30\times0.25)^{0.6}}{(100)^{0.4}(0.02)^{0.3}} + \left(\frac{30}{0.3} \times \frac{1}{60}\right)$	
$= 11.88 + 1.67 = 13.55 \approx 14$ minutes	

b. Major flood

$t_c = t_{sheet\ flow} + t_{channel\ flow}$
$= \frac{6.94(30\times0.25)^{0.6}}{(150)^{0.4}(0.02)^{0.3}} + \left(\frac{30}{0.3} \times \frac{1}{60}\right)$
$= 10.11 + 1.67 = 11.78 \approx 12$ minutes

Design rainfall intensities (interpolating values stated in Table 4.6).

Bioretention basin	Minor flood		Major flood	
	T_c (minute)	Rainfall intensity (mm/hr)	T_c (minute)	Rainfall intensity (mm/hr)
	14	127.3	12	231.3

Using the rational method (Eq. 2.18a), the obtained peak flows are:

Minor flood	$Q = \frac{\left(0.75 \times \frac{127.3}{1000} \times 1230\right)}{360} = 0.33 \text{ m}^3/\text{s}$
Major flood	$Q = \frac{\left(0.94 \times \frac{231.3}{1000} \times 1230\right)}{360} = 0.74 \text{ m}^3/\text{s}$

Step 3: *Design inflow systems* for scour protection, coarse sediment forebay, and street hydraulics. *Inlet scours protection* using rock beaching, or rock works has been placed to restrict flows entering the curb opening. Stormwater runoff is conveyed to the bioretention basin through coarse sediment forebay. The required volume of forebay sediment storage as per Eq. 4.5:

here,

Adjacent allotments area (ha) = 0.1230 ha

Capture or removal efficiency = 80% (assume)

Sediment loading rate = 1.6 m^3/ha/year

Frequency of cleanout = at least once in every two years

$$V_s = 0.123 \times 0.8 \times 1.6 \times 2 = 0.315 \text{ m}^3$$

Coarse sediment forebay area computed as the volume is dividing by the depth. Usually, the required depth is within 0.3 m below the filter media top unless any other available local/regional recommendations. Thus, the area is $\left(\frac{0.315}{0.3+0.3}\right)$ Or 0.525 m^2.

This area needs to be checked the capturing capacity of the 1mm and above particles using Eq. 4.6.

Here,

For the three-month flow rate, assume the wet season, i.e., Q_{3mnt}, which is usually 0.5 times of Q_1.

$$t_c = t_{sheet\ flow} + t_{channel\ flow}$$
$$= \frac{6.94(30 \times 0.25)^{0.6}}{(85.5)^{0.4}(0.02)^{0.3}} + \left(\frac{30}{0.3} \times \frac{1}{60}\right) = 12.65 + 1.67 = 14.32 \approx 14 \text{ minutes}$$

So, I_1 = 92.58 mm/hr

$$Q_1 = \frac{\left(0.75 \times \frac{92.58}{1000} \times 1230\right)}{360} = 0.24 \text{ m}^3/\text{s}$$
$$Q_{3mnt} = 0.5 \times 0.24 = 0.12\ m^3/sec$$

Then,

$$R = 1 - \left[1 + \frac{1}{0.5} \cdot \frac{0.1}{0.12/0.525}\right]^{-0.5} = 0.7312 \text{ i.e., } 73\% \text{ of 1 mm particles}$$

Streetscape application—size curb opening: the depth and width of the gutter need to be determined at the curb opening for defining the hydraulic head. To feature flow conditions through the curb opening, and the length of the opening can be determined applying a broad crested weir equation (Eq. 4.4):

$Q_{weir} = 0.33$ m ^{3}/s [minor flood]
$C_w =$ Weir coefficient, 1.66
$L =$ Length of opening (m)
h = Depth of water above the weir (pit) crest = 100 mm (assume)
Using the information mentioned above:

$$0.33 = 1.66 \times L \times (0.1)^{3/2} \approx 6.3 \text{ m}$$

Therefore, implementing a 6.3 m long opening allows gutter flow depth and width; they remain unchanged for the upstream of the curb opening.

Step 4: Material specification of the bioretention filter and drainage layer in the perforated under-drainage system. Slot width of 1.5 mm within a perforated pipe has been selected to reduce the sand wash-out risk. A three-layer filter media is required to be provided comprising 600 mm filter media of sandy loam, a 100 mm transition layer of coarse sand, and a 200 mm drainage layer of fine gravel.

Step 5: *Design* underdrain and conveyance capacity checks (optional) to ensure horizontal flow through perforated pipes in the drainage layer to minimize drainage flow within the filter media.

where

A = Area of bio retention basin = 1230 m^2
K = Filter layer saturated hydraulic conductivity (extracted from Table 2A.2) = 0.18 m/hr
h_{max} = Depth of pondage above the filter = 0.3 m
d_{filter} = Depth of filter media = 0.6 m
Using Eq. 4.7, the maximum filtration rate:

$$Q_{max} = \frac{(0.18 \times 1230)}{3600} \times \frac{0.3 + 0.6}{0.6} = 0.0922 \text{ m}^3/\text{s}$$

Perforations Inflow Check

Like bioretention swale, the sub-surface drainage system's inlet capacity is required to restrain choke in the system. Conventionally, half of the total holes in a standard perforated pipe are assumed to be blocked. Consider as an orifice, and the flow rate should be estimated applying Eq. 4.3as:

$$\text{Head above pipe, } \text{h} = 0.95 \text{ m}[0.6 \text{ m(filter depth)}$$

$$+ 0.3\ \text{m(max.pond level)} + 0.05\text{(half of pipe diameter)}]$$

Assume sub-surface drains with 50% pipes blocked:
Clear opening = 2100 mm^2/m
Blocked openings (i.e. 50% of the clear opening) = 1050 mm^2/m
Slot width = 1.5 mm
Slot length = 7.5 mm
Number of rows = 6
Diameter = 100 mm
Number of slots per m = $\frac{1050}{1.5 \times 7.5} = 93.3$
Assume orifice flow conditions, the Cd = 0.61.
Inlet capacity/m of pipe:

$$\begin{aligned} Q_{perf} &= \left[0.61 \times (0.0015 \times 0.0075 \times 93.3)\sqrt{2 \times 9.81 \times 0.95}\right] \\ &= 0.0028\ \text{m}^3/\text{s} \end{aligned}$$

Inlet capacity per m × total length [to convey runoff from 3 bioretention basins, 2 parallel pipes @ 100 m pipe (proposed)]

$$\begin{aligned} &= 0.0028 \times (2 \times 100) \\ &= 0.56\ \text{m}^3/\text{s} >> 0.0922\ \text{m}^3/\text{s} \end{aligned}$$

Perforated Pipe Capacity

The flow rate in the perforated pipes is estimated to ensure the conveyance capacity of the maximum filtration rate. Two perforated pipes of 100 mm diameter parallelly placed at 2% grade towards the overflow pit. Assume Manning's n as 0.02, and using Manning's equation, the flow rate is:

$$Q_{flow\ per\ pipe} = \frac{1}{0.02} \times \frac{\pi}{4}(0.1)^2 \times 0.025^{2/3} \times 2^{1/2} = 0.0474\ \text{m}^3/\text{s}$$

Then,
Q_{total} = 0.0948 m^3/s (for two pipes) > 0.0922 m^3/s, and hence OK.

Step 6: The *impermeable lining* as filter media is not required. In the catchment, the surrounding soils are clay to silty clays with a K_{sat} of approximately 3.6 mm/hr. Then, sandy loam media (K_{sat} of 180 mm/hr) is proposed as the filter media.

Step 7: *Sizing of the overflow pit* depends on the bioretention basin requirements. The minimum permissible head is 500 mm over the pit crest to facilitate minor flood disposal. Design consideration of the overflow pit is either submerged or free-flowing condition. Under the free-flowing condition, a broad crested weir equation determines the required weir length (i.e., Eq. 4.4). Submerged outlet condition uses an orifice equation to estimate the critical area between openings in the grate cover (i.e., Eq. 4.3).

For, $Q = 0.33\ m^3/s$, $B = 0.5$, $C_w = 1.66$ and $h = 0.5$ m.
Using Eq. 4.4:

$$0.33 = 0.5 \times 1.66 \times L \times 0.5^{3/2}$$

Then, L = 1.127 m of weir length required.
For drowned conditions, using Eq. 4.3:

$$0.33 = 0.5 \times 0.61 \times A \times \sqrt{2 \times 9.81 \times 0.5}$$

Then, $A = 0.345\ m^2$ (equivalent to 588 mm × 588 mm pit).

Hence, free overflow conditions dominate, and the pit needs to be greater than 588 mm × 588 mm.

Step 8: *The selection of the vegetation* should be based on regional guidance and consultation by landscape architects. Thus, both the treatment facilities and the landscape of the area could be achieved.

Step 9: Verification checks with the scour velocities over the vegetation and the achieved water quality after passing the bioretention basin. Potential scour velocities of bioretention swale should be less than 0.5 m/s and 2.0 m/s for minor flood and major flood discharge.

4.5.5 Permeable Pavements

Permeable paving is a method to enable infiltration of stormwater runoff, applicable to the paving vehicle and pedestrian pathways. Thus, there are two approaches for stormwater management using permeable pavement, i.e., increased infiltrations to reduce runoff, seep into the pavement, and infiltrate the underlying soils reducing runoff at a site. A typical cross-section of permeable paving has shown in Fig. 4.13, comprises (i) permeable pavers, (ii) permeable joint material, (iii) sand (instead of open-graded bedding course), (iv) gravel (instead of open-graded subbase reservoir), (v) coarse gravel/retention trench (instead of open-graded subbase reservoir), (vi) geotextile fabric.

Generally, pervious concrete applies as permeable paving material; also, porous asphalt, interlocking concrete pavers, concrete grid pavers, and gravel or grass-filled plastic reinforced grids are widely used. This type of pavement comprises multiple layers to facilitate surface protection.

- *Pervious concrete* omits the fine aggregates in the conventional concrete mix, i.e., Portland cement, coarse aggregate or gravel, and water. There is a void content of 15–35% due to removing the finer particles (i.e., sand) from the concrete mix. Thus, water infiltrates the underlying soil rather than surface inundation or surface runoff. Pervious concrete commonly applies in sidewalks and parking lots.

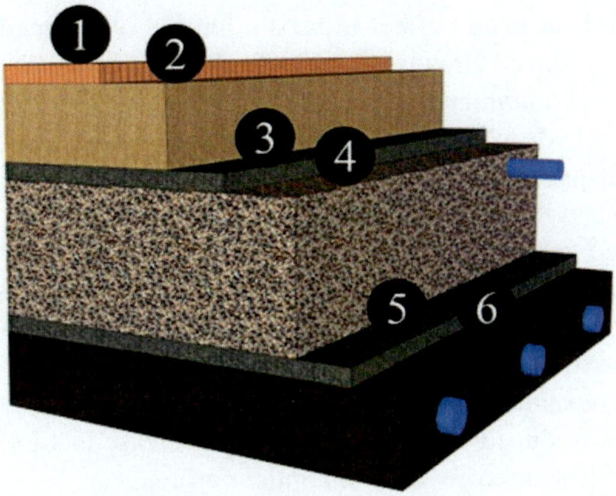

(1) Permeable pavers, (2) Sand (instead of open-graded bedding course), (3) Geotextile fabric, (4) Coarse gravel/retention trench (instead of open-graded subbase reservoir), (5) Geotextile fabric, and (6) Filter layer

Fig. 4.13 Permeable pavements

- *Porous asphalt* is a standard mixture that uses less fine aggregate than conventional asphalt. The bituminous binder binds both fine and coarse aggregate, maintaining a void content of 15% to 35%. Although this has a rougher texture, the surface appearance is conventional, and the surface layer is usually thinner. Porous asphalt is generally used in pedestrian pathways, i.e., greenways and low volume, low-speed vehicular traffic applications, i.e., parking lots, curbside, parking lanes on roads, and residential or side streets.
- *Pavers* are made from different materials, for example, Permeable Interlocking Concrete Pavers (PICP), Permeable Interlocking Clay Brick Pavers (PICBP), Concrete Grid Pavers (CGP), and Plastic Turf Reinforcing Grids (PTRG). PICPs are organized dense concrete blocks towards a pattern form with filling fine aggregate between the pavers. PICBPs use bricks as a replacement for concrete. Using concrete CGPs having larger openings or apertures configured in lattice-style, the gaps are filled with soil or grass. Similarly, using plastic units, PTRG increases structural strength to topsoil and reduces compaction. In both cases, i.e., CGPs and PTRG, infiltration could be improved by vegetation and increase the underlying soil's permeability.

Advantages

- *The natural hydrological cycle* could be achieved through increased stormwater infiltration, and, after storm events, lower streamflow levels contribute to the adjacent waterbody. Additionally, increased stream base flow promotes more groundwater recharge. Evaporation from the permeable pavement offers a cooler surface and reduces the heat island effect in urban. On the other hand, stream stability increases due to reduced stream velocities and peak flows.

- *Standing water is eliminated* by permeable pavements. Thus, it improves vehicle braking, reduces hydroplaning[2] on roadways, and reduces resistance to freezing and thawing effects. Similarly, it contributes to the city's wellbeing and improves urban plantations with greater access to water and air.
- *Both surface and groundwater conditioning* as the materials for permeable pavement retain fine soluble particulate, nutrients, sediments, heavy metals, and other pollutants from stormwater runoff. It also minimizes road salt (for winter de-icing) applications due to improved drainage conditions.
- *The permeable pavement could offer aquatic habitat-friendly conditions* through stream stability, reduced thermal pollution, decreased chloride levels in the receiving waters (in case of road salt application).

Design Considerations

Permeable pavement design considers traffic loads, storm volume storage, draining duration, and water quality requirements. Thus, design consideration requires for both structural and hydraulic components. The structural design process of permeable pavement needs to determine the pavement thickness and the underlying layers. The retention trench is designed to capture the desired stormwater volume produced from the design storm. Historical time series on storm duration and return periods are required to obtain a design storm.

The *minimum depth method* estimates the depth of the retention trench for a given porous paved area. The *minimum area method* computes the required porous pavement surface area for a given design depth of the retention trench.

Following steps are required for the design consideration:

Step 1: *Confirmation of proper permeable pavement* and available preliminary treatment.

Step 2: *Design flow determination* for permeable paving should consider major and minor storms.

Step 3: *Porous paving system sizing* for a particular design storm event. A selected porous paving system should be able to capture and infiltrate this specific design storm event. In Australia, this is usually 3-month and 1-year ARI design storms (DPI, IMEA and BCC 1992). The required 'detention volume' (V_d) of a permeable paving system is the difference between inflow and outflow volumes during the storm duration. The *inflow volume* for the design storm on the porous paving system (treatment surface) could be calculated as per Argue (2007):

$$V_i = \frac{A_s I}{10^3 \times 360} D \tag{4.8}$$

[2] Hydroplaning occurs when a tire encounters more water than it can scatter.

where,

V_i = Inflow volume (m^3)

A_s = Permeable pavement surface area (m^2)

D = Duration of the storm (hr)

Total runoff is produced from rainfall onto the porous paving system and from additional impervious areas. The inflow (V_i) is the product of the design storm flow and the storm duration:

$$V_i = Q_d D \tag{4.9}$$

Here

Q_d = Design storm for sizing (rational method, $Q = \frac{CIA}{360}\left(m^3/s\right)$.

The *permeable paving system outflow* through the base or sides of the filter media depends on the available area and ponding depth. This outflow is calculated based on the filtration rate for a storm event.

Fully drained conditions, i.e., no detention depth above the permeable pavement surface $\left[i.e. \frac{h_{max}+d}{d} = 1\right]$ in Eq. 4.7, then the outflow volume is:

$$V_0 = Q_{\max} D \tag{4.10}$$

Then, the required detention volume (V_d) of permeable pavement is:

$$V_d = \frac{V_i - V_0}{n} \tag{4.11}$$

Here, n is the porosity of the retention trench. Values are available in Table 2.1; for gravel, this value is 0.35.

Thus, the surface area (A_s) of the permeable paving system could be checked using Argue's (2007) equation:

$$A_s = \frac{Q_P}{(1 - B).K_{sat}} \tag{4.12}$$

Here

Q_p = Peak inflow on the permeable paving surface (m^3/s)

B = Blockage factor (based on non-porous structural elements, e.g., plastic/concrete grids)

K_{sat} = Saturated hydraulic conductivity of paving surface, e.g., concrete/asphalt or permeable material between pavers

Step 4: *Underdrainage design and check*

The perforated pipes should undergo several checks to ensure adequate size:

- The slots are sufficient to convey the maximum filtration rate;
- The pipe capacity is greater than the maximum filtration rate. The flow capacity of the perforations is as per the basic orifice formula, i.e.:

$$Q_{perf} = B.C_D.A_o\sqrt{2gh} \tag{4.13}$$

where

Q_{perf} = Flow-through the perforations (m^3/s)
C_D = Discharge coefficient for orifice, 0.6 to 0.65 (Medaugh and Johnson 1940)
A_o = Area of the orifice (m^2)
h = Maximum depth of water above the pipe (m)
B = Blockage factor (suggested for 50% blockage of the perforations)

- The filter materials restrict to be washed out through the perforated pipes by providing a transition layer.

Step 5: *Emptying time*

Emptying time through the filter layer is obtained as the volume of water in temporary storage divided by the filtration rate. The volume of water in temporary storage is defined by the product of the storage dimension and porosity. The filtration rate is calculated as the product of hydraulic conductivity and infiltration area, A_{inf}. Thus, the emptying time (t_e):

$$t_e = \frac{1000.V_d.n}{A_{inf}.K_{sat}} \tag{4.14}$$

The stored stormwater generally exfiltrate within 24–48 h following a storm event. Typically within 12–72 h, to obtain required storage for successive storm events. However, this should be selected considering land uses and land cover; for instance, the recommended range is up to 84 h in Australia (Engineers Australia 2006). The soil infiltration capacity for stormwater depends on different soil types, and *Table 2.1* (stated in Chap. 2) considers during design.

Step 6: *Requirement for impermeable lining.*

The ground rule is if the K_{sat} of the permeable paving system's saturated hydraulic conductivity is *ten* times more than that of the surrounding in-situ soil profile, the impermeable lining is not required.

Step 7: *Elements of the porous paving layers.*

Essential elements are porous paving surface, retention/aggregate layer, and geofabric. The suggested stone/gravel size is between 25 and 100 mm diameter, and for geofabric the recommended minimum perforation or mesh is 0.25 mm.

Step 8: *Sizing overflow pit/pipe.*

Using the typical pipe equations and the inlet and outlet conditions and friction losses, overflow pipes are determined within the piped. Generally, the pipe or culvert operates under outlet control considering the submerged condition. Friction losses are negligible for short pipe connections, and these should be computed using Manning's equation.

An *overflow pit* could be designed for either submerged or free-flowing conditions using an orifice equation and a broad crested weir equation, respectively.

Example problem 4.3 With an improved water supply system, the seven-storied building stated in Example problem 3.3 becomes self-sufficient but has faced difficulties in recent years with stagnant waters during the wet season. The available catchment area distribution is the roof, paved and unpaved areas are 234.1 m^2, 105.91 m^2 and 20.90 m^2, respectively.

Step 1: *Confirmation of proper porous pavement* and available preliminary treatment.

Step 2: *Design flow for* the hydraulic design of permeable paving considers both major and minor floods, i.e., for 10 years and 100 years ARI peak flows

Overland sheet flow component through paved area = 21.34 m [parallel to the building]

The sheet flow path is predominately concrete surfaces, with a longitudinal slope of 2% and a typical Manning's n = 0.015

Minor flood

$T_c = t_{sheet\ flow}$ Application of kinematic wave equation (Eq. 2.21) for sheet flow computation

$$= \frac{6.94(L.n)^{0.6}}{I^{0.4}S^{0.3}} = \frac{6.94(21.34\times 0.015)^{0.6}}{(100)^{0.4}(0.02)^{0.3}} = 1.79 \approx 2\ minutes$$

Major flood

$$T_c = \frac{6.94(21.34 \times 0.015)^{0.6}}{(150)^{0.4}(0.02)^{0.3}} = 1.53 \approx 2\ minutes$$

Design rainfall intensity (extrapolating values stated in Table 4.7) for T_c = 2 minutes is 155.7 mm/hr and 264.3 mm/hr for minor and major floods.

Table 4.7 Rainfall – runoff details

		1 year	10 years	100 years	Remarks
Average recurrence interval (ARI) rainfall (mm/hr)		85.5	100	150	Based on the regional IDF curve
rainfall intensity (mm/hr) for different time of concentration, T_c	T_c = 20 minute	88	120	200.5	
	T_c = 15 minute	90.6	129.5	219	
	T_c = 10 minute	100.5	140.5	239.5	
	Tc = 5 minute	120	150	255	
Runoff coefficient	Roof	0.75	0.75	1.19	Table 2.3
	Flat lawn, 2%	0.05	0.05	0.13	
	Asphaltic and concrete	0.70	0.70	1.19	

Using the rational method (Eq. 2.16a), the obtained peak flows are:

Minor flood	Major flood
$Q = \frac{\left(0.70\times\frac{155.7}{1000}\times105.91\right)}{360} = 0.032\ \mathrm{m^3/s}$	$Q = \frac{\left(1.19\times\frac{264.3}{1000}\times105.91\right)}{360} = 0.092\ \mathrm{m^3/s}$

Step 3: *Porous paving system sizing* for a specific design storm event. A selected porous paving system should be able to capture and infiltrate this particular design storm event. In Australia, this is usually 3-months and 1-year ARI design storms (Engineers Australia 2006).

Three-month flow rate, assume the wet season, i.e., $Q_{3mnt,}$ and usually, this is 0.5 times of Q_1.

$$t_c = \frac{6.94(21.34 \times 0.015)^{0.6}}{(85.5)^{0.4}(0.02)^{0.3}} = 1.91 \approx 2 \text{ minutes}$$

So, $I_1 = 131.7$ mm/hr (extrapolating Table 4.7), and the duration is 30 minutes (assumed following Article 2.4.3).

$$Q_1 = \frac{\left(0.75 \times \frac{131.7}{1000} \times 105.91\right)}{360} = 0.029\ \mathrm{m^3/s}$$
$$Q_{3mnt} = 0.5 \times 0.029 = 0.014\ m^3/sec$$

The *inflow volume* for the design storm on the permeable pavements as per Eq. 4.7 is:

$$V_i = \frac{A_s I}{10^3 \times 360} D = \frac{105.91 \times 131.7}{10^3 \times 360} \times \frac{30}{60} = 0.19\ \mathrm{m^3}$$

The outflow volume is calculated as:

$$V_0 = Q_{max} D = 0.014 \times \frac{30}{60} = 0.007\ m^3$$

Then the required detention volume (V_d) is:

$$V_d = \frac{0.19 - 0.007}{0.35} = 0.523\ m^3$$

Here, n is the porosity of the retention trench = 0.35.

The *depth of the porous paving system* is 60 mm (Assume). Thus, a surface area of 8.71 m^2 could be designed.

The surface area of the permeable paving system needs to be checked using the equation Eq. 4.12. Here, B and K_{sat} are 50% and for gravel 0.05 m/s respectively. Then,

$$A_s = \frac{0.19}{(1 - 0.5) \times 0.05} = 7.6 \text{ m}^2$$

So, the designed area is more than the requirement.

The rest of the design would then be continued following steps 4–9 with the bioretention area or bioretention basin design experience.

4.5.6 Rain Barrels or Cisterns

Rain barrels or *cisterns* are the widely practiced onsite rooftop runoff management due to the cost-effectiveness and ease of maintaining as a retention and detention device. A rain barrel system includes a catchment, guttering and downspouts, screens or filters, and a rain barrel (Fig. 4.14). In other words, this is the rainwater storage with only preliminary treatment as described in *Sect. 3.6*, Chap. 3. Additionally, stored water from the rain barrel is accessible through a tap or faucet, and a hosepipe facilitates an irrigation system. The planning for scale-wise desired catchment stormwater management control strategy would be different. For instance, *watershed-scale rain*

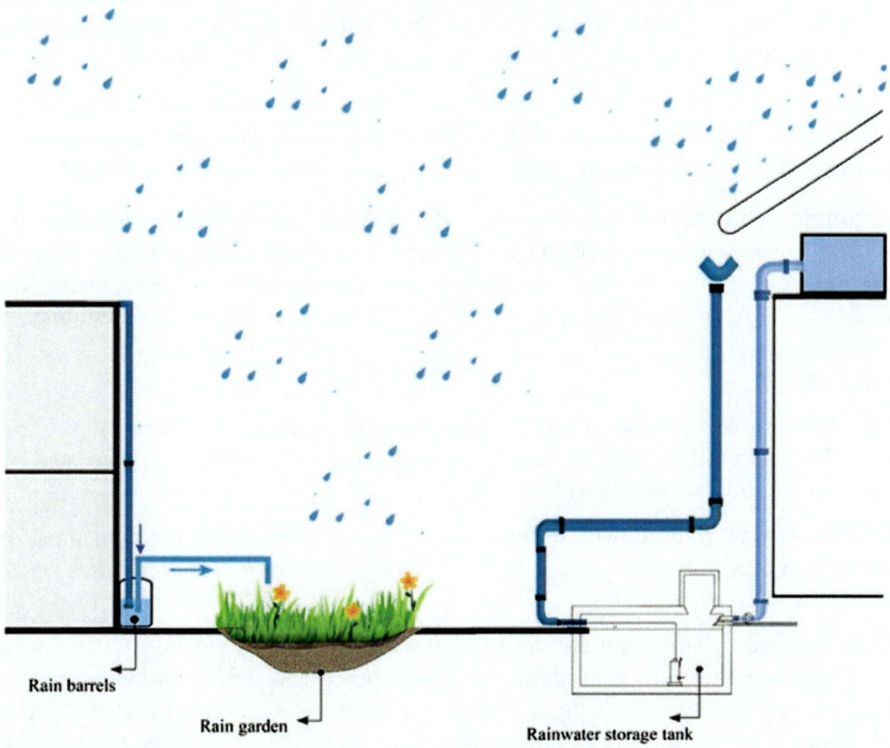

Fig. 4.14 Rain barrel

barrels are highly owner-dependent, allowing discharges to designated infiltration areas within the property before the storm. On the other hand, *site-scale rain barrels are usually based on rooftop runoff management and the number of barrels* planned to serve the desired downpipes.

Following steps are required for the design consideration:

Step 1: Pre-storage treatment should be ensured by confirming pre-filtration and first flushing as Sects. 3.6.1 and 3.6.2 in Chap. 3.

Step 2: There is a thumb rule, i.e., 1 mm of rainfall on 1 m^2 of catchment produces 1 liter of water to store. Sizing rain barrels should be conducted following the *Tank Sizing* described in Sect. 3.3.2 following available manufactures' configurations.

Although a rainwater harvesting system might fail to reduce other stormwater management systems' requirements, if on-site stormwater management is required, the rainwater harvesting system needs to revise the size to lower storage and revise the difficulties of other lot-level infiltration systems. Rainwater overflow handling systems based on the overflow discharge locations are as follows:

- Rainwater tanks are located above grade or below grade practice discharge to grade via either gravity flow or pump assistance for discharge towards the grade.
- Storm overflows disposed of by gravity flow into a storm sewer regardless of rain tanks location, whether above or below grade;
- Rainwater tanks situated below ground, on-site infiltration of rainwater overflows rather than pumping to grade. The overflow drainage piping fails to connect the storm sewer, then discharges to the soakaway pit via gravity flow.

The stormwater management techniques include infiltration basins, soak ways, infiltration trenches, dry swales, cisterns, tree box filters, curbless roads with swales downspout disconnection, etc., often face site constraints as described in Table 4.8.

Example problem 4.4 Following the Example problem 4.3, the owner plans to remain with a rooftop rainwater harvesting system and beautify the building. Suggest (a) the required space for the *rain garden* in this building. (b) for other technologies, if the nearby open space just opposite to roadside turns to a two-storied building.

Solution:

(a) The available, paved area for the seven-storied building is 105.91 m^2 (Fig. 4.15). For every 50 m^2 area of runoff, the rain garden should be 1 m suggested by Melbourne water (2009).

Thus, letting some allowances for boundary wall at both sides of the front unpaved areas each of 3.35 m × 1.5 m requires converting into a rain garden of 2 m^2, i.e. (2 m × 1 m) with 1 m depth.

(b) Rainwater harvesting for stormwater management also extends to the city or community level. Considering owners' choice and the required stormwater management suggests the technical requirements. Regarding the future population growth and LULC, the stormwater management, as shown in Fig. 4.16, could be comprised of:

Table 4.8 Comparison of site constraints for stormwater management (Rossman 2006)

Stormwater management practice	Depth to high water table or bedrock (m)[1]	A typical ratio of the impervious drainage area to the treatment facility area	Native soil infiltration rate (mm/hr)[3]	Head (m)[4]	Space (%)[5]	Slope (%)[6]	Pollution hotspots[7]	Setbacks[8]
Green roof	Not applicable	1:1	Not applicable	0	0	0	Yes	None
Roof downspout disconnection	Not applicable	[5–100 m^2][2]	Amend if <15 mm/hr[9]	0.5	5–20	1–5	Yes	B
Soakway, infiltration trench or chamber	1	5:1–20:1	Not a constraint	1–2	0–1	< 15	No	B, U, T, W
Bioretention	1	5:1–15:1	Underdrain required if <15 mm/hr	1–2	5–20	0–2	No	B, U, W
Biofilter (filtration only bioretention design)	Not applicable	5:1	Not applicable	1–2	2–5	0–2	Yes	B,T
Enhanced grass swale	1	5:1–10:1	Not applicable	1–3	5–15	0.5–6	No	B, U
Vegetated filter strip	1	5:1	Amend if <15 mm/hr[9]	0–1	15–20	1–5	No	None
Permeable pavement	1	1:1–1.2:1	Underdrain required if <15 mm/hr	0.5–1	0	1–5	No	U,W
Dry swale	1	5:1–15:1	Underdrain required if <15 mm/hr	1–3	5–15	0.5–6	No	B, U,W
Rain barrel	Not applicable	[5–50 m^2][2]	Not applicable	1	0	NA	Yes	None
Cistern	1	[5–3000 m^2][2]	Not applicable	1–2	0–1	NA	Yes	U,T
Perforated pipe system	1	5:1–10:1	Not a constraint	1–3	0	<15%	No	B, U, T,W

Notes:

[1] Minimum depth between the base of the facility and the elevation of the seasonally high water table or top of bed rock.

[2] Values for rain barrels, cisterns, and roof downspout disconnection represent typical ranges for impervious drainage area treated.

[3] Infiltration rate estimates based on measurements of hydraulic conductivity under field saturated conditions at the proposed location and depth of the practice.

[4] The vertical distance between the inlet and outlet.

Table 4.8 (continued)

[5] Percent of open pervious land on the site.
[6] Slope at the stormwater management technique.
[7] Suitable in pollution hotspots or runoff source areas where land uses or activities have the potential to generate highly contaminated runoff (e.g., vehicle fuelling, servicing or demolition areas, outdoor storage or handling areas for hazardous materials and some heavy industry sites).
[8] Setback codes: B = Building foundation, U = underground utilities, T = trees, W = Drinking water wellhead protection areas.
[9] Native soils should be titled and amended with compost to improve infiltration rate, moisture retention capacity, and fertility.

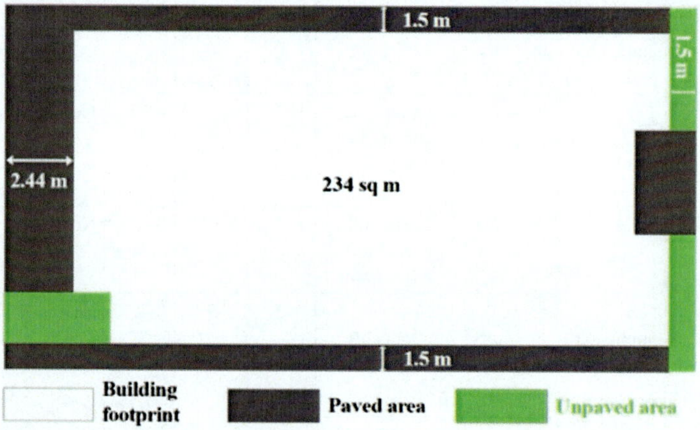

Fig. 4.15 Example problem 4.4

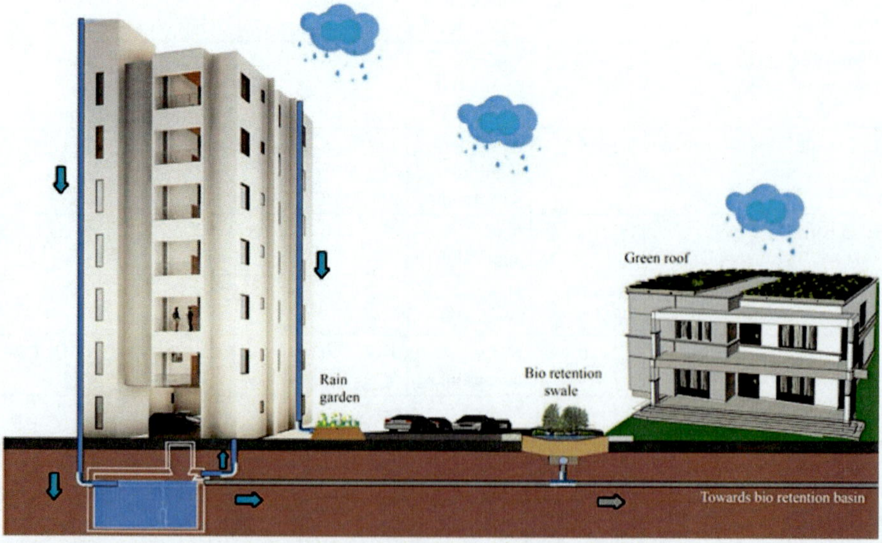

Fig. 4.16 Rainwater harvesting system for stormwater management

- Green roof on a two-storied building,
- Rain garden in front of the seven-storied building,
- Bio-retention swale between the road and footpath, both sides of the road contributes to bio-retention basin, and
- All over open spaces with the building compound would have permeable pavements.

4.6 Design and Installation Guidelines

Although technology-wise design and installments are described in respective sections, the overall design process includes the following steps:

Step1: Location identification;

Step 2: Determination of the adopted stormwater management techniques;

Step 3: Define the site details;

Step 4: Outline catchment development, conceptual earthworks, and drainage layout;

Step 5: Confirming rainwater tank requirements;

Step 6: Design stormwater treatment measures and check with *Steps 2–5*.

4.7 Management Strategies

Following management strategies should be adopted:

- Lack of maintenance results in premature green roof failure, and the management strategies should consider meeting the issues described in *Sect. 4.5.1* in this Chapter;
- Regular inspection and maintenance to avoid underdrains clogging (as described in *Sect. 4.5.2*) are required for the rain gardens to achieve the desired infiltration of stormwater;
- Bioretention swales and basins
  - Remove sediment, trash, or other organic material as needed;
  - Monthly inspect banks and surrounding drainage areas, including out parcels and parking lots, are required to reduce erosion and sedimentation;
  - Perimeter mowing (maintain a 100–150 mm height);
  - Inspect plants and replace as necessary;
  - Inspect for proper drawdown/ clogging;
  - Mulch renewal and replacement.
- Permeable pavements
  - Inspection requires after storms to ensure proper drainage; stagnant water/ponding should be disposed of by 36 h.

  - Vacuum sweep to get rid of sediment on the pavement. The frequency of sweeping might be once or twice a year.
  - For cold countries, care requires removing snow from the pavement surface.
  - The structural integrity of the permeable pavement should undergo routine inspection for required actions on repair or replacement.
  - Drain outfall inspection.

- Rain barrels or cisterns

  - Gutters should be cleaned regularly or installing cover to prevent clogging during water collection.
  - Insect populations should be reduced using required mess;
  - Avoid overflowing barrels;
  - Hold an annual cleaning inside the barrel, and examine the barrel outside for traces of cracks.
  - Care should be taken for rain barrels in countries that experience snowfall to avoid cracks in the barrel.

While stormwater management works considering rainwater storage tanks:

- Suppose the overflow drainage piping discharges above grade. In that case, overflow examines erosion, and a coarse screen at the end of the overflow drainage pipe restricts the passage of dirt and debris. This should be inspected annually.
- If the overflow pipe discharges below grade, routine inspection should be carried out to handle blocked or poorly performing overflow drainage systems. These signs belong to water damage to the rainwater tank and associated components, water leaking, and water backing up rainwater inlet lines and top-up drainage piping.
- During the inspection, cleaning, or repairing of components of the overflow system, all the required safety precautions should be followed.

References

Alberta Environmental Protection (1994) Best management practices and spill response : guidance document. Industrial Wastewater Branch Air and Water Approvals Division Alberta Environmental Protection. Available at: https://open.alberta.ca/publications/best-management-practices-and-spill-response-guidance-document#summary.

Argue JR (ed) (2007) Water sensitive urban design: basic procedures for 'source control' of stormwater. University of South Australia, AWA.

Bob B et al. (2012) Rain garden guide. https://raingardens.info/

Burian SJ, Edwards FG (2002) Historical perspectives of urban drainage. In: 9th International conference on urban drainage, pp 1–16. https://doi.org/10.1061/40644(2002)284

Burton GA, Pitt RE (2002) Stormwater effects handbook: a tool for watershed managers, scientists and engineers. CRC Press LLC. Lewis Publishers, New York.

Chang N-B et al (2018) Global policy analysis of low impact development for stormwater management in urban regions. Land Use Policy 70:368–383. https://doi.org/10.1016/j.landusepol.2017.11.024

DPI, IMEA and BCC (1992) Queensland urban drainage manual (QUDM), prepared by Neville Jones & Associates and Australian Water Engineering for DPI, IMEA & BCC, Brisbane.

Durrans SR (2003) Stormwater conveyance modeling and design. Haestad Press, Waterbury, CT.

Engineers Australia (2006) Australian runoff quality, engineers Australia. ACT. http://www.arq.org.au/

Fair GM, Geyer JC (1954) Water supply and waste disposal, vol 2. Wiley, New York.

Fardin HF et al. (2014) Sanitation and water management in ancient South Asia. In: AngelakisAN, Rose JB (eds) Evolution of sanitation and wastewater technologies through the centuries. IWA Publishing, pp 43–53

Fletcher TD et al. (2015) SUDS, LID, BMPs, WSUD and more—the evolution and application of terminology surrounding urban drainage. Urban Water J 12(7):525–542, Taylor & Francis. https://doi.org/10.1080/1573062X.2014.916314

Grigg NS (2012) Water, wastewater, and stormwater infrastructure management. CRC Press, pp 365.

Hoyer J et al. (2011) Water sensitive urban design: principles and inspiration for sustainable stormwater management in the city of the future. Jovis, Hamburg, Germany.

James R (2012) Modelling LIDs using PCSWMM and EPA SWMM5. https://www.pcswmm.com/Downloads/USEPASWMM

Medaugh FW, Johnson GD (1940) Investigation of the discharge and coefficient of small circular orifices. Civil Eng 422–424.

Medl A et al (2017) Vertical greening systems – a review on recent technologies and research advancement. Build Environ 125:227–239. https://doi.org/10.1016/j.buildenv.2017.08.054

Melbourne water (2009) Rain gardens. https://www.melbournewater.com.au/community-and-education/help-protectenvironment/raingardens. Accessed 3 Oct 2017.

Oberts GL et al (1989) The water quality performance of select urban runoff treatment systems. Metropolitan Council, Publication No. 590-89-062a, p 170.

Rossman LA (2006) Storm water management model quality assurance report, dynamic wave flow routing, EPA/600/R-06/097.

Saraswat C et al (2016) Assessment of stormwater runoff management practices and governance under climate change and urbanization: an analysis of Bangkok, Hanoi and Tokyo. Environ Sci Policy 64:101–117. https://doi.org/10.1016/j.envsci.2016.06.018

USEPA (1983) Results of the nationwide urban runoff program, PB84-185552. Washington, DC.

USEPA (1999) Preliminary data summary of urban storm water best management practices. Washington D.C: ingineering and analysis division/office of science and technology. United States Environmental Protection Agency, p 448. http://www.epa.gov/waterscience/guide/stormwater

USEPA (2016) National menu of best management practices (BMPs) for stormwater. National pollutant discharge elimination system.

Chapter 5
Greywater Water Reuse

5.1 Introduction

Greywater, which results from human activities, i.e., wastewater, has been used once and then considered reuse with treatment. In recent years, significant developments have been recorded worldwide on greywater reuse practices and policies to address the water smart city. Due to the diversity of water use, the quantity and compositions of the generated greywater vary from country to country. This chapter presents the usual greywater reuses, knowledge on the typical quality and quantity of the generated greywater, computation of the greywater footprint, existing codes, integrated uses with either rainwater or/and stormwater, general consideration on design and installation, management strategies, and worked out example problems. Thus, knowledge on conventional to extensive treatment might be enriched. Ensuring 'zero waste' for the water smart city is essential despite the land scarcity and expensive treatment facilities. Considering socio-economic, health issues, and environmental factors to facilitate the greywater reuse practices, a summary of the practiced management strategies is presented. Integrated rainwater-greywater or stormwater-greywater might reduce the treatment costs compared to only greywater reuse. Describing the integrated approach in this chapter attempted to bridge the cost-effective method of greywater treatment and the managed aquifer recharge.

5.2 Greywater Reuse

Pure water includes spring water, rainwater, and groundwater. *Blackwater* is a term used to refer to water containing feces, urine, toilet water, and toilet paper. *On the other hand, greywater* is the water that results from human activities like dishwashing, laundry, and bathing (Fig. 5.1). Therefore, greywater is all the water produced from the house except toilet water. Kitchen sinks and dishwashers generated greywater containing bacteria, and the legal barrier restricts using them as recycled greywater.

© The Author(s), under exclusive license to Springer Nature Switzerland AG 2022

A. Akter, *Rainwater Harvesting—Building a Water Smart City*, Springer Water,
https://doi.org/10.1007/978-3-030-94643-2_5

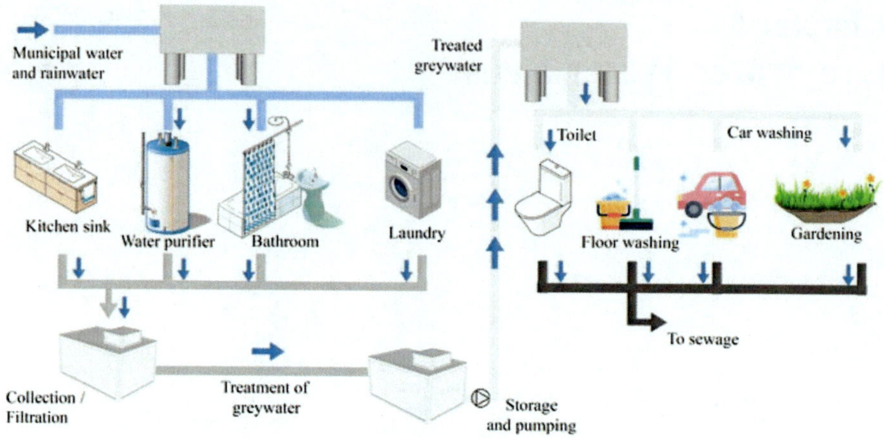

Fig. 5.1 Greywater recycling

Around 3000 years ago, the ancient civilizations of Babylonia and Assyria started combined greywater and stormwater sewage systems. In recent years, significant developments are observed in greywater reuse practices and policies. Worldwide different expressions have been practicing, i.e., 'Greywater' or 'Grey water' in the UK (Otterpohl et al. 1999; Jefferson et al. 2001; Queensland Government 2008), 'Gray water' or 'Graywater' in the US (Del Porto and Steinfeld 2000; Wilderer 2004)) and 'foul water' in Hong Kong. Similarly, researchers also termed 'Reclaimed water' (Crook and Surampalli 1996; Gregory et al. 1996), 'Diluted wastewater' and 'Light wastewater' (Ledin et al. 2001). Further characterization is found in research (Wilderer 2004): Wastewater containing feces is *Brown water*, and wastewater containing urine is *Yellow water*. Many researchers consider adding kitchen water to greywater. This combination causes organic loads to double, increasing treatment cost, operation, and maintenance requirements. The advancement of reuse has been observed in water-conscious countries. The components of a typical greywater system are as follow (Fig. 5.1).

- Greywater *source* (household wastewater without toilet wastes);
- *Collection system* comprises of drains in bathroom, laundry and kitchen water;
- *Treatment system* built in the underground of the buildings;
- *Storage facilities* provided for potable water storage after the treatment process, and
- *Greywater reuse* is the treated greywater sent to the toilet or washing machine, but never equivalent to potable water.

Zero waste is a resource management approach through reduction, reuse, and recycle. These systems reduce greenhouse gases by saving energy and eliminating the need for landfills and incinerators.

Greywater systems vary significantly based on the generated quantity and quality. These are the water recycling (i) without purification, i.e., production of non-potable water, and (ii) after purification for domestic tasks. Further classifications are based on the type of filtration or treatment (BSI 2010).

(a) Direct reuse systems (without treatment)

Collection of greywater from appliances using simple devices and supply directly to the Points of Use (POU), without any or minimal treatment, and with or without storage, e.g., using a greywater diverter valve. This system is usually applied to garden irrigation.

(b) Short retention systems

Short retention approaches are generally practices for garden irrigation. In this system, essential filtration or treatment techniques are engaged to skim debris off the collected greywater surface and permit sedimentation to the tank's bottom. These systems overlook odour, and water quality issues as the treated water allow storage only for a short duration.

(c) Basic physical/chemical systems

A filter is usually assigned to exclude debris after collection before greywater storage in a physical system. In due course, chemical disinfectants (e.g., chlorine or bromine) are applied to disrupt bacterial growth. These systems are used for greywater reuse in garden irrigation and toilet flushing.

(d) Biological systems

These are based on either aerobic or anaerobic digestion to eliminate unwanted organic material in the collected greywater. During aerobic treatment, aeration of water is conducted using pumps or aquatic plants.

(e) Bio-mechanical systems

The advanced domestic greywater reuse system, as shown in Fig. 5.1, combines biological and physical treatment to remove organic matter by microbial cultures and settles solid. They inspire bacterial activity by aerating greywater.

(f) Hybrid systems

These combine more than one of the systems mentioned above.

5.3 Advantages and Disadvantages of the Greywater Reuse

Limited available urban spaces face difficulties offering proper treatment and greywater reuse facilities; however, this issue carried off using greywater reuse technologies requiring small spaces. Treated greywater can be reused in urban areas for toilet flushing and reinforcing the concept of urban agriculture/gardening.

Advantages

The potential ecological advantages include:

- freshwater extraction lowering from rivers and aquifers;
- topsoil nitrification;
- lessen the use of toxic chemicals;
- groundwater recharge;
- increase soil tilth and fertility;
- reduce landfill;
- increased plant growth, support flourishing landscapes;
- encourage other resource-conserving practices.

Management advantage

- extend the life of septic leach fields and reduce pumping costs;
- conserve water and lower water bills;
- provide better purification than septic or sewer;
- improve climate safety and lower energy use.

Disadvantages

- the whole system is expensive, including installment, operational, and maintenance, compared to the recycled water cost;
- health problems are one of the most significant concerns while greywater reuses;
- places with the absence of a water meter mean no payment by the customers for water consumption (i.e., the non-revenue water), thus, lack of financial incentive to install a greywater system;
- traditionally the reliability of greywater systems are very low;
- inadequate water in the sewer system to pass the waste causing fouling;
- greywater reuse as near-potable requires clearance from the regional health department; and
- currently, regulations on reused water quality are absent except for nationwide initiatives.

5.4 Quantity of Greywater

Greywater makes up around 50–80% of total wastewater from residential or commercial buildings. Many factors could influence the generated wastewater: the potable water demand, consumption patterns, urban growth, geographical location, household category, plumbing fixtures/fittings type, and age (Schuetze and Santiago-Fandiño 2013; Samayamanthula et al. 2019). Countrywide the daily average generated greywater per capita extends up to 110 liters (Table 5.1). To calculate the quantity of greywater, generally, three components are considered i.e.

Table 5.1 Estimated volumes of greywater based on the available literature

Wastewater source	Volume (liter/person/day)									
	Australia(Fane and Reardon 2013)	UK(Cameron et al. 2014)	US(Jungels et al. 2013)	Israel (Friedler 2004)	Netherlands (Hernandez-Soriano and Jimenez-Lopez 2012)	Germany(Nolde 2000)	Denmark(Eriksson et al. 2002)	Switzerland (Morel and Diener 2006)	Vietnam (Busser et al. 2006)	Nepal (Shrestha et al. 2001)
Shower	63	28	38	55	40	20–40	50	52	30–60	–
Hand basin	6	17	41		23	20–40	10			–
Kitchen tap	12	13	14	30	9	4–6	25	28	15–20	–
Dishwasher	5									–
Laundry tap	2	13	20	13	3	3–10		30	15–30	–
Washing machine	13					10–15	8			–
Total—grey water	101	61	103	98	74	57–111	93	110	80–110	72

a. Net water supplied, by deducting the amounts of water losses in the networks and percentile of sewage network connection, as well as the amount of water used in agriculture in each desired land;
b. Population dataset for existing and projected demand calculation;
c. Generated greywater quantity could be up to 70% of household water consumption (WHO 2006).

Thus,

$$\begin{aligned} Estimated\ greywater\ quantity &= (net\ water\ supply - network\ water\ loss) \\ &\quad \times (1 - \%\ of\ sewage\ connection) \\ &\quad \times population \times 0.70 \times 0.001 \times 365\ days \end{aligned} \tag{5.1}$$

The quantity of greywater has been generated as a function of the household dynamics. The influencing factors are the existing water supply and infrastructure, inhabitants/users per household, users' age distribution, lifestyle, water consumption patterns, etc.

5.5 Greywater Quality

Compared to black water, the main characteristic of greywater contains low organic matter and nutrients, i.e., Nitrogen, Potassium, and Microbes. As for the greywater content of heavy elements, this would be similar to black water. Generally, the greywater content of microbes and pathogens is less than black water. However, viruses, bacteria, parasites, and intestinal worms might exist in greywater. Sources wise their water quality characteristics are described in Table 5.2.

The quality of greywater varies, considerably including sources and installations, for the following reasons:

- The quality of greywater is influenced by geographical locations, i.e., rural and urban areas, also by demographics and level of occupancy;
- The difference in the quality of water used in the house due to difference in sources, viz., tap water, or groundwater or rainwater;
- Variation in lifestyle, patterns of water consumption, and use of chemical products among households. The higher the consumption rate resulted, the better the quality of greywater due to the reduction in pollutants concentration;
- Type of distribution network for drinking water;
- The quality of greywater varies according to climate or season;
- The physical condition of greywater collection pipe network.

Based on the available greywater quality, the collected or treated greywater can be reused in urban areas for toilet flushing and urban agriculture/gardening. Worldwide

Table 5.2 Characteristics of untreated greywater (Wright 1996; Queensland Government 2008)

Water source	Characteristics
Laundry	Microbiological: Variable thermotolerant coliform loads Chemical: Soaps and soiled clothes cause a high concentration of Na, PO_4, Boron (B), detergents, NH_4, and Total Nitrogen (TN) Physical: Higher suspended solids and turbidity produce grey colour in water as well as high temperature Biological: Higher BOD
Bathtub and shower	Microbiological: Lower thermotolerant coliform loads Chemical: Soap, shampoo, and other substances cause high Na, PO_4, B, detergents, NH_4, and TN Physical: Higher suspended solids, turbidity, and temperature Biological: Lower BOD
Kitchen	Microbiological: Variable thermotolerant coliform loads Chemical: High concentration of Na, PO_4, B, detergents, NH_4, and TN Physical: Food particles, oils, fats, grease resulting from leftovers and turbidity
Evaporative cooler	Salinity
Swimming pool	Chlorine and salinity

different places have set different permissible levels for the application of greywater (Table 5.3).

5.6 Greywater Treatment System

The greywater treatment is conducted through separation, biological treatment, and disinfection. The *separation system* detached the solids after the greywater collection and released them to the drain. The household's *biological treatment* is usually conducted through Bio-Matter Resequencing Converter (BMRC), which comprises three tanks, i.e., a surge tank, aeration tank, and clarifying tank. *Disinfection* of *potable water system* destroys the containing microorganisms after biological treatment by application of ozone or UV. Among them, biological treatment is the primary approach to degrade organic compounds. A cost analysis is essential to decide a suitable greywater treatment system identification. The following parameters should be considered:

- Connections
  - Main water line (optional)
  - Service water (optional)

Table 5.3 Permissible greywater quality from the available literatures

Place	Application	BOD_5 (mg/l)	TSS (mg/l)	Turbidity (NTU)	Faecal coliform (Counts/100 mL)	Total coliform (Counts/100 mL)	pH
Japan (Tajima 2005)	Toilet flushing	10	–	5	<10	<10	6–9
Germany (Al-Jayyousi 2003)	Wastewater reuse	20	–	1–2	500	100	6–9
Israel (Gross et al. 2007)	Wastewater reuse	10	10	–	<1	–	
Canary wetland, Spain (USEPA 2004)	Wastewater reuse	10	3	2	–	2.2	
California, US (USEPA 2004)	Unrestricted water reuse	10	–	2	–	2.2 (Average) 23 (maximum)	6–9
Florida, US (USEPA 2004)	Unrestricted water reuse	20	5	–	<240	–	
Queensland, Australia (Queensland Government 2008)	Greywater reuse for garden watering	20	30	2	<4	100	
British Columbia, Canada (CMHC 2004)	Unrestricted urban water reuse	10	5	2	2.2	–	

  - Input to pipe unit (usually **n**ominal **d**iameter DN 100[1])
  - Overflow to sewage (usually DN 100)
  - Air-ventilated working room, with necessary floor drains

- Control unit
  Aeration time and demand-based pre-programmed filtration
- Main water line
- Solenoid valve (optional, DN 13).

5.6.1 Separation of Greywater

Plastic pipes (PVC) are usually used to separate greywater from black water at the household level. The pipelines are extended to a collection point where treatment occurs. Therefore, the direction of the sewage pipes in the house should be changed, which might be in some cases, involve the removal of existent tiles. The basic requirements of water separation pipes are:

- Greywater drainage pipe diameters should follow the amount of water to be treated;
- Valves are installed to regulate and control the greywater flow; and
- Sufficient slope to facilitate the flow of greywater.

5.6.2 Greywater Treatment Technologies

Greywater treatment technologies include three main stages, and these are physical, biological, and tertiary treatment.

First Stage: Physical Treatment

At this stage, collected greywater stores in a sedimentation tank allow sediments to deposit to the bottom of the tank. The lightweight material and solids, i.e., oils and foams, remain floating. The next stage includes filtration, sedimentation, and coagulation processes (Fig. 5.2a). The advanced physical treatments include coarser sand, soil, and membrane filtration with or without disinfection (Fig. 5.2b and c).

Second Stage: Biological Treatment

In the second stage of treatment, biodegradable organic matter (both soluble and suspended) are removed in addition to suspended solids. Conventionally, this treatment stage includes sedimentation and screening, disinfection, and passing over the biological rector (Fig. 5.3a). In the advanced technology, greywater input passes

[1] ISO 6708 (ISO 1995) defines the nominal size as DN, DN followed by a dimensionless whole number that is related to the physical size of the bore or outside diameter of the end connections, expressed in mm.

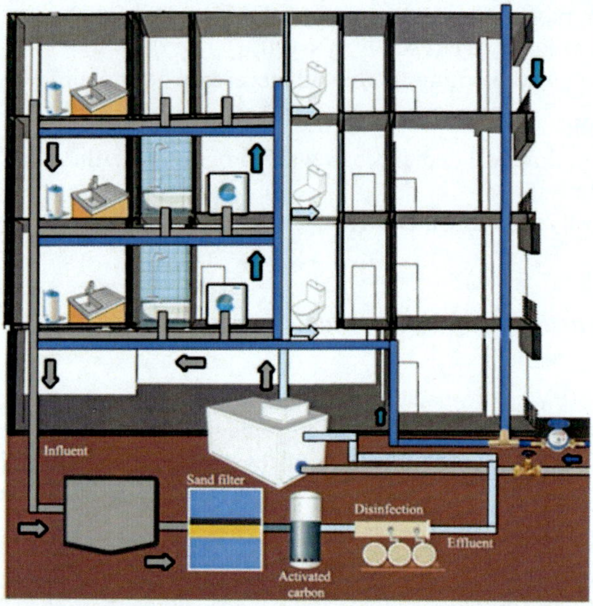

(a) Treated greywater for non-potable use

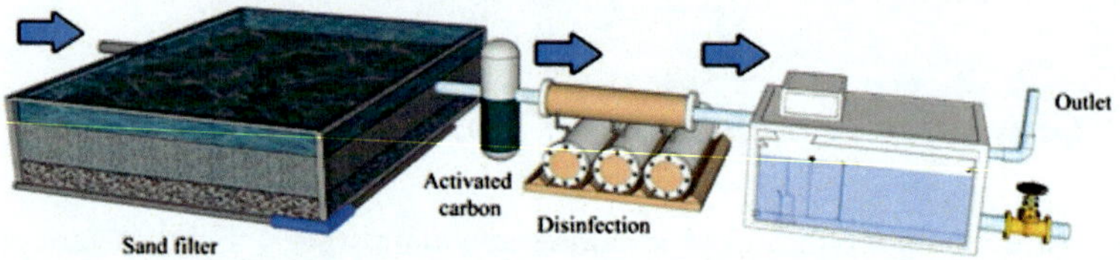

(b) Conventional treatment setup

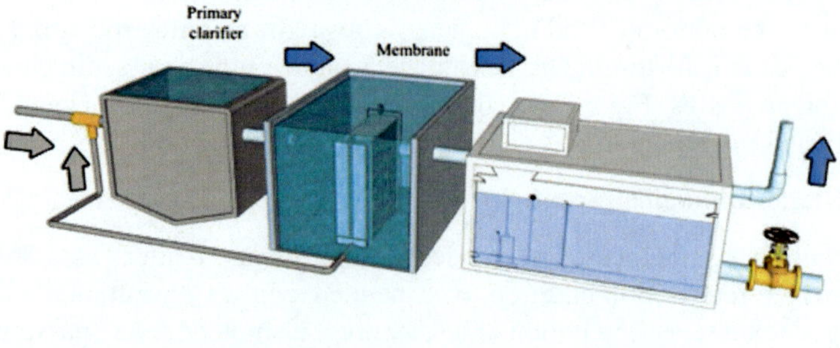

(c) Advanced treatment setup

Fig. 5.2 **a**, **b** and **c**: Physical treatment system

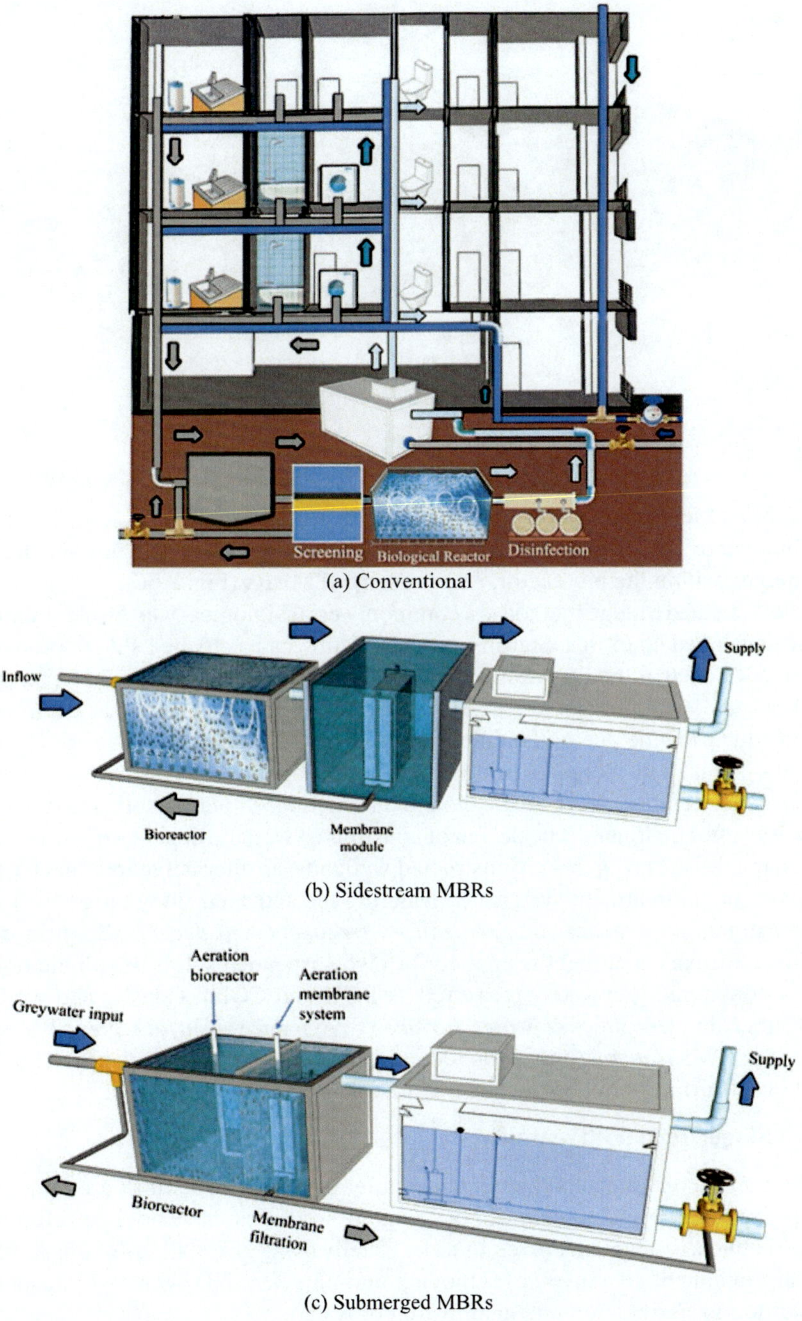

(a) Conventional

(b) Sidestream MBRs

(c) Submerged MBRs

Fig. 5.3 Biological technologies **a** conventional **b** side stream MBRs and **c** submerged MBRs

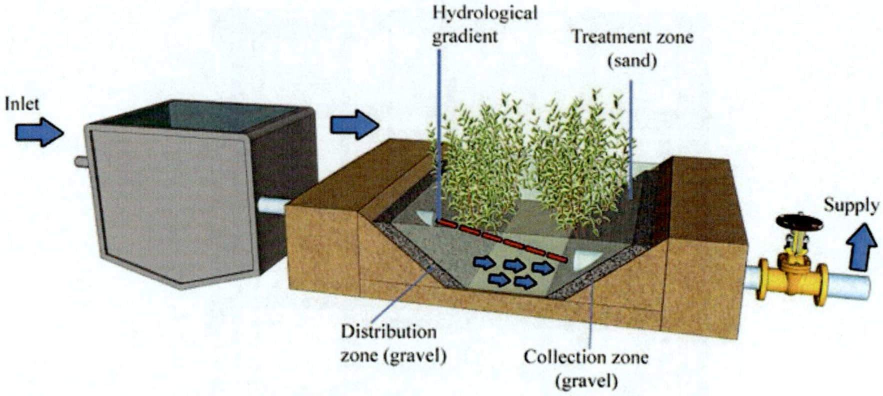

Fig. 5.4 Typical extensive technology

through the Membrane Bioreactor (MBRs) either the membrane module place after the bioreactor, i.e., the sidestream MBRs (Fig. 5.3b) or the membrane module submerged within the bioreactor, i.e., submerged MBRs (Fig. 5.3c).

The Activated Sludge method is a commonly used biological wastewater treatment in the developed countries. Starting in 1913, a full-scale activated sludge process by Arden & Lockett at the Davyhulme sewage treatment works in Manchester (Ardern and Lockett 1915) placed a milestone. Afterward, the basic concept has been widely practiced worldwide due to the unique flexibility of operation.

Natural treatment methods include stabilization lagoons and constructed wetlands. *Constructed wetlands* work on the biological functions of plants, soil, and organisms for wastewater treatment. The design of constructed wetland depends on the wastewater characteristics (Fig. 5.4). Constructed wetlands, an alternative treatment system to conventional treatment systems considering the activated sludge process, follow two treatment approaches, i.e., *centralized treatment* and *decentralized treatment* methods. Primary treatment is applied for high suspended solids or soluble organic matter containing greywater (measured as BOD and COD). Constructed wetlands are further divided into *free water surface* (FWS) and *subsurface flow*. The water surface in FWS is maintained 100–500 mm above the built wetlands bed. The water level in subsurface flow remains below the constructed wetlands bed.

Third Stage: Tertiary Treatment

At this stage, contaminants that remained after the previous settings are removed to comply with reuse specifications. For example, solids that have not been eliminated in the secondary treatment are removed by usually using gravel filters or micro-filters. Tertiary treatment also involves removing nutrients, i.e., Nitrogen and Phosphorus, in addition to sterilization and disinfection of water.

Other Treatments

Simple technologies are two-stage systems based on sedimentation to eliminate larger solids followed by disinfection. Usually, this treated greywater applies to subsurface irrigation. For example, the rules and regulations of Western Australia permit the reuse.

Chemical treatment systems are based on coagulation with aluminum. This system is the combination of (i) coagulation, sand filter, and Granular Activated Carbon (GAC), (ii) electro-coagulation with disinfection, and (iii) photocatalytic oxidation with titanium dioxide and UV. These two technologies could achieve acceptable water quality within shorter contact times.

Soak pit is a closed, porous-walled chamber that often permits to soak water into the ground slowly. The organic-rich greywater in the septic tank needs to be filtered out. Pre-settled effluent from the septic tank disposes to the underground chamber towards infiltrates into the surrounding soil.

A greywater infiltration system is a relatively new concept that uses local facilities. An example could be drawn from the Djenné city in Mali, with its approximately Twenty thousand inhabitants, located in the inner delta of the Niger River. Greywater-related problems were mitigated through the local greywater infiltration in 2002 using local material and labour. By 2004, the greywater infiltration system became popular. Thus, greywater could contribute to groundwater recharge through natural treatment.

Common greywater treatments are usually expensive; however, there are also some advantages for the consumers. Both advantages and disadvantages are described in Table 5.4.

5.7 Greywater Footprint (GWF)

The greywater footprint (GWF) indicates the required volume of water to assimilate a pollutant load entering a waterbody. The Water Footprint Assessment Manual (Chapagain et al. 2011) and Intergovernmental Panel on Climate Change (IPCC 2008) recommend a three-tier approach to estimate GWF. *Tier 1*: uses a leaching-runoff fraction to record the applied chemical substance on the soil to evaluate their significance in groundwater or surface water system. *Tier 2*: relates simple and standard models considering the topographic, hydro-meteorological, and soil characteristics of the site where the chemical substance is applied. *Tier 3*: uses sophisticated modeling techniques. This approach is relatively expensive due to the involvement of data-intensive physically-based models and higher computation time. Thus, the GWF is the pollutant load entering a water body (in mass/time) by dividing the critical load times (Lcrit, in mass/time) the water body flow (in volume/time).

$$GWF = \frac{\text{Pollutant load}}{L_{crit}} \times (waterbody\ flow)\left[\frac{Volume}{time}\right] \tag{5.2}$$

Table 5.4 Common greywater treatments

Treatment technique	Description	Advantages	Disadvantages
Sand filter	Beds are of sands/coarse/mulch to trap and absorb contaminants as greywater flows through	• Simple operation • Low maintenance and operation costs	• High capital cost • Reduces pathogens but fails to eliminate • Clogged and flooded if overloaded
Membrane bioreactor	Combined aerobic biological treatment and filtration to eliminate pathogens	• Highly practical with proper design and operation	• High capital cost • High operation cost • Complex operational requirements
Activated carbon filter	Activated carbon has been treated with oxygen to open up millions of tiny pores between the carbon atoms	• Simple operation • reduces both organic and inorganic chemicals	• High capital cost • Fail to remove Sodium, Nitrate, etc • Stop working while the bonding sites are filled
Disinfection	Chlorine, ozone, or ultraviolet light can be used to disinfect greywater	• Highly effective in eliminating bacteria • Low operator skill requirement	• Chlorine and ozone can produce toxic by-products • Ozone and ultraviolet can be affected by the organic contents
Aerobic biological treatment	Air is bubbled to transfer oxygen from the perspective into the greywater	• A high degree of operations flexibility for accommodating the treated water	• High capital cost • High operation cost • Complex operational requirements

The critical load (Lcrit) is the pollutants load that entirely consumes the assimilation capacity of the receiving water body. Lcrit is calculated by multiplying the flow (in volume/time) by the difference between the ambient water quality standard of the pollutant (the maximum permissible concentration c_{max}, in mass/volume) and its natural concentration in the receiving water body (i.e., c_{nat}, in mass/volume).

$$L_{crit} = (waterbody\ flow) \times (c_{\max} - c_{\text{nat}})\left[\frac{\text{mass}}{\text{time}}\right] \tag{5.3}$$

By inserting Eq. 5.3 in 5.2:

$$GWF = \frac{\text{Pollutant load}}{c_{\max} - c_{\text{nat}}}\left[\frac{\text{Volume}}{\text{time}}\right] \tag{5.4}$$

5.8 Greywater Codes

The available greywater codes are either performance-based or prescriptive-based. *Performance-based codes* are highlighted health and safety necessities for greywater recycling systems. These codes are straightforward and often fail to specify the required detailing to ignore pooling and runoff. *Prescriptive codes* select the necessary building materials and the component of a greywater system. Innovations and regulations advanced the early history of greywater about worldwide greywater reuse, then widely rejected. Earlier initiation of indoor plumbing, greywater reuses for irrigation, and mishandling often combines greywater with blackwater. Due to severe health hazards, greywater recycling is expelled by plumbing codes. In 1989 Santa Barbara County, the first US jurisdiction, permitted greywater irrigation. Followed by this, Santa Barbara launched the world's first plant and soil biocompatible laundry detergent in 1990. Also, worldwide greywater reuses is regulated by law in many countries and states in the US, Australia, Japan, Germany, Cyprus, Saudi Arabia, Oman, and Jordan. These regulations cover potential greywater reuses for irrigation, indoor usage, and heat reclamation (Table 5.5). However, considering possible health risks, greywater reuses in many countries for indoor are still under consideration, for instance, in the US, Canada, etc.

Table 5.5 Worldwide available guidelines and codes for greywater reuse

Country	Code/policy	Contribution to the knowledge
Australia	i. Australian guidelines for water recycling (NRMMC-EPHC 2006) ii. Interim NSW guidelines for the management of private recycled water schemes (NSW 2008) iii. Code of Practice for the Use of Greywater in Western Australia (Department of Health 2010)	i. Possible health risks ii. Risk management approach iii. Minimum requirements for the reuse (single residential, domestic premises)
Canada	Canadian Guidelines for Domestic Reclaimed Water Ottawa (Health Canada 2010)	Describe health effects due to greywater reuse and management framework
UK	Greywater systems (BSI 2010)	Bathroom greywater reuse for non-potable purpose
US	i. Arizona (ADEQ 2002) ii. California (State of California 2011) iii. Texas water law (TWDB 2005)	• Subsurface drip irrigation with greywater • Greywater use efficiency in building
Germany	Betriebs-Fachvereinigung and Darmstadt (2005)	Greywater recycling requirements for planning, execution, operation, and maintenance of greywater recycling plants

Example problems 5.1 As described in Example problem 3.3, For a seven-storied building, the ground floor is dedicated to parking, and 32 occupants reside in 6 apartments. The daily per capita water demand is 120 liters. They have planned to adopt water-saving devices and water-saving appliances with the following guidelines:

Shower: If each person has one time 7.5 min shower per day through water-saving devices (7.5 lpm)

Bathtub: 96 liters per person per week

Clothes Washing: per-household three times/week through a water-saving front loading washing machine, and each time requires 40L = 120 liters per week.

Dishwasher (when it is full) = 476 liters per week

Teeth brush twice a day =100 ml of water/day

Hand wash three times a day = 300 ml of water/ day

How much water could be conserved?

Solution:

With water-saving showerhead = $32 \times 1 \times 7.5 \times 7.5 \times 7 = 12{,}600$ liters per week.

Bathtub use = $32 \times 1 \times 96 = 3072$ liters per week.

Clothes Washing = $120 \times 6 = 720$ liters per week.

Kitchen consumption: For a week, if vegetables were rinsed in a shallow sink of water using a water-saving kitchen tap, similarly, the dishes undergo rinsed before entering the dishwasher, and the dishwasher only ran once it is full $476 \times 6 = 2856$ liter per week.

Hand basin: for teeth brush and handwashing = $\left(\frac{300}{1000} \times 32\right) \times 7 = 67.2$ liters per week.

The total average weekly usage for this building using water-saving devices and water-saving appliances is:

$= 12600 + 3072 + 720 + 2856 + 67.2 = 19,315.2$ litres per week.

On the other hand, with their existing practices, the weekly usage for this building is:

$= 120 \times 32 \times 7 = 26,880$ liters per week.

Thus, using water-saving devices and water-saving appliances in this building, the greywater generation could be reduced up to 28%.

5.9 Combined Greywater and Rainwater Harvesting

Individual greywater reuse systems face difficulties with inadequate volume to meet the required non-potable use and charge more for treatment. A combination with the rainwater harvesting system might offer a feasible solution. Before integration, a thorough assessment should be made of each system individually to identify their performances to meet the desired demand. The integrated systems can either be operated as separate, independent systems (Fig. 5.5) or be combined into a single

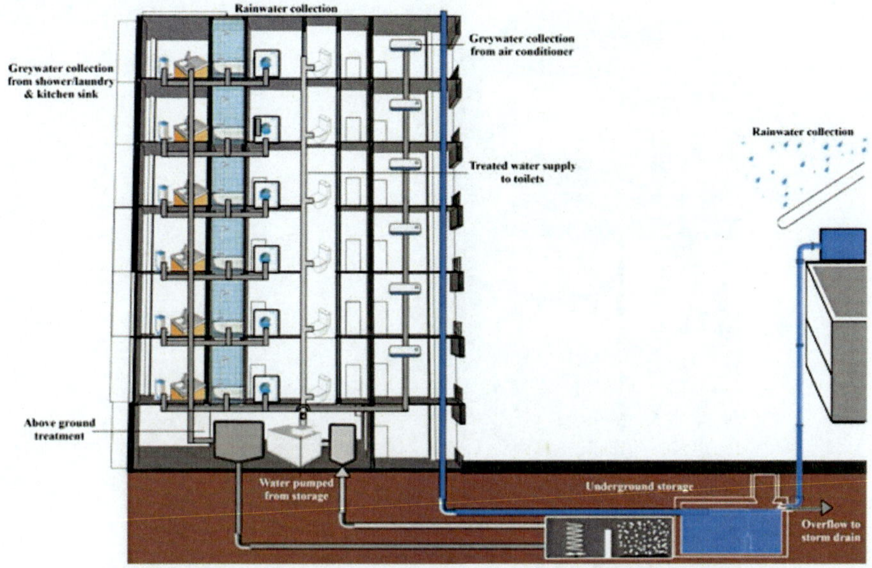

Greywater components

- Collection pipework carrying bathroom greywater for treatment
- Greywater treatment unit (both primary and secondary)
- Distribution pipework carrying treated greywater to storage
- Distribution pipework convey treated greywater to POU

Rainwater components

- Gutter
- Rainwater storage
- Distribution pipework convey treated greywater to POU

Fig. 5.5 Integrated greywater recycle and rainwater harvesting system with separate storage and supply

supply source (Figs. 5.6 and 5.7). Untreated greywater and rainwater are stored together; afterward, the overflows or bypass provisions discharge into sewerage. Also, excess rain is diverted to a soakaway or drainage system if the storage is full, thus minimizing filthy drains. Care should be given to all aspects of the treatment system instalment considering the following issues:

(a) Local facilities should be consulted to select overflow and bypass connections for surface water drainage;
(b) The plumbing fittings and fixtures at the point of integration should meet the regional standard. For example, BS 8525-1 for Great Britain (BSI 2010); and
(c) If the integrated systems are comprised of different manufacturer's products, the compatibility of the systems needs to be investigated for proper action.

Systems integrated before treatment are subjected to:

i. Influent variability;
ii. Required treatment equipment;

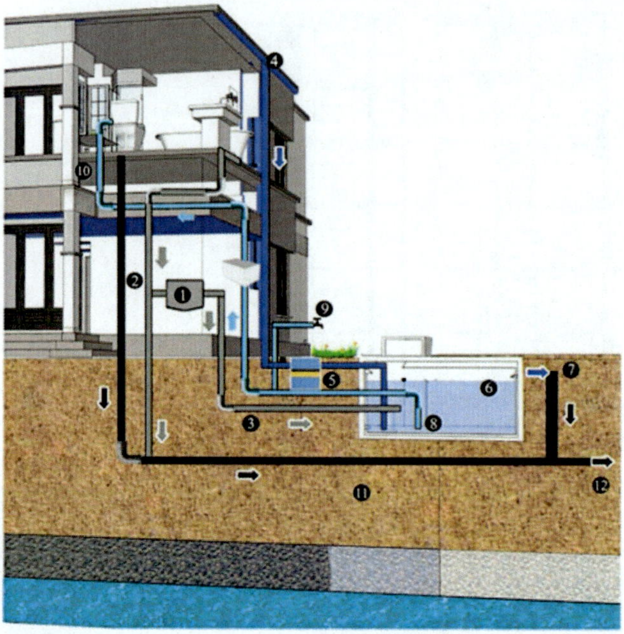

Greywater components
1. Greywater treatment unit
2. Pipework collects bathroom greywater for treatment
3. Distribution pipework convey treated greywater to storage

Rainwater components
4. Gutter
5. Filter

Shared components
6. Storage tank
7. Overflow pipe
8. Calmed inlet
9. Garden tap
10. Distribution pipework carrying rainwater/ treated greywater to POU
11. Foul drain
12. Foul sewer

Fig. 5.6 Integrated greywater and rainwater harvesting system with single storage and direct supply

iii. Identify the management strategy for handling excess water loads;
iv. Environmental considerations for electricity consumption; and
v. Treatment validity of the greywater and rainwater.

After treatment, greywater and rainwater systems may be combined at various points, e.g., tank/cistern, distribution pipework, or at the POU, without (Fig. 5.5) or with (Fig. 5.6) direct supply connection. Greywater and rainwater harvesting systems are integrated after treatment with indirect non-potable supply systems.

The *demand ratio* (for non-potable water and availability) of greywater and rainwater should be estimated for designing the integrated system.

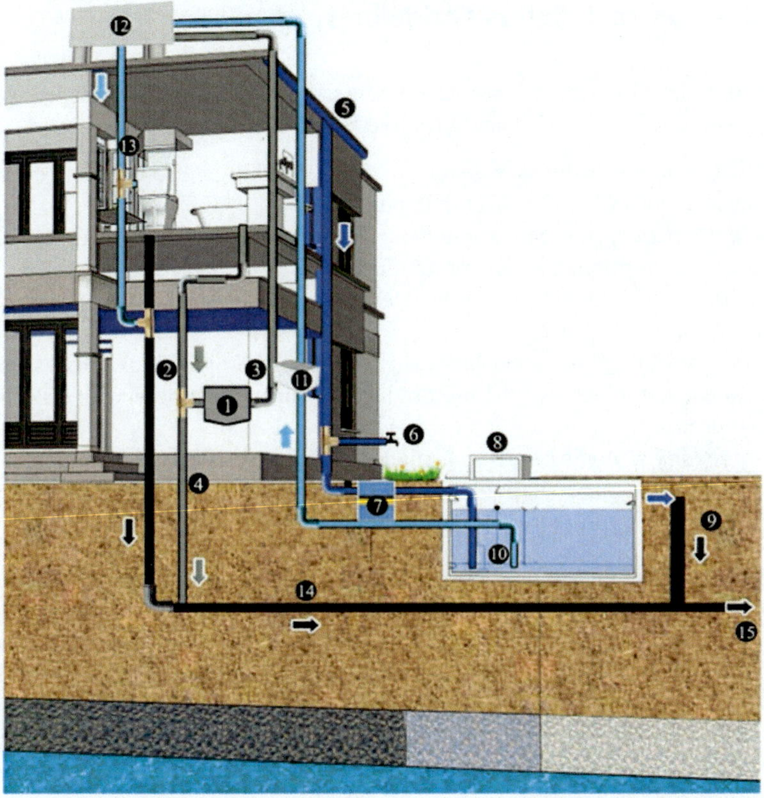

Greywater components
1. Greywater treatment unit
2. Pipework collects bathroom greywater for treatment
3. Distribution pipework carrying treated greywater to storage
4. Bypass

Rainwater components
5. Gutter
6. Bypass/ Garden tap
7. Filter
8. Storage tank
9. Overflow pipe
10. Calmed inlet
11. Distribution pipework carrying rainwater

Shared components
12. Storge cistern
13. Distribution pipework carrying rainwater / treated greywater to the POU
14. Foul drain
15. Foul sewer

Fig. 5.7 Integrated greywater and rainwater harvesting system with indirect non-potable supply (systems integrated after treatment)

5.10 Design and Installation Guidelines

Although technique-wise design and instalments were described in respective sections, the overall design process includes five steps:

- Rain*water is potentially harvestable* from roofs, parking lots, adjacent open spaces, and vertical walls of the high-rise building.
- *Potable water consumption and greywater generation* depend on the freshwater source, their consumption, and deserving reuses.
- *Characterization of rainwater and greywater* are needed to design the necessary treatment for non-potable uses.
- *Selection and design of treatment processes* consider the operations, processes, and systems needed to achieve the required treatment to meet deserving criteria for the final use.
- *Water storage tank sizing* is based on demand, availability of rainwater and greywater, and available space (individually or combined in a group of buildings, walkways, and parking lots) for tank instalment.

Example problem 5.2 The ground floor is dedicated to parking for a seven-storied building, and 32 occupants resident six apartments. The available roof footprint is 234.1 m^2. There is a two-storied building just opposite this building while it crosses the main road. The two-storied building has nine occupants with an available roof footprint of 234 m^2. The daily per capita water demand is 120 liters. The mean annual rainfall is recorded as 2918.1 mm. In a newly developed residential area, both of these two buildings owners are planning for an integrated rainwater and greywater management (Fig. 5.8) as below:

Fig. 5.8 Integrated greywater and rainwater harvesting

- Seven storied buildings will reuse the greywater in association with the collected rainwater, and
- Two storied buildings will contribute both rainwater and greywater to the centralized traffic island garden.

Quantify the amounts of available nonpotable water?

Step 1: Rainwater potentially harvestable

The contributing catchment area needs to be calculated to estimate rainfall volume. Then, the conveyance of collected rainwater and greywater. Using Table 2.3 and 3.1, the yield coefficient for concrete pavement and flat roof as 0.7 and 0.8, respectively.

For the *seven-storied building*, annually available rainwater as per the Example problem 3.3.

Option I: Rooftop only = (546.50)/365 = 1.5 m^3/day i.e., 45m^3/mean month.

Option II: All Possible sources [roof, paved area and 50% of a verticle wall] = (546.50 + 216.30 + 677.83)/365 = 3.95 m^3/day i.e., 118.5 m^3/(mean month).

For the *two-storied building*, annually available rainwater.

a. Roof = $234.1 \times 2.9181 \times 0.8 = 546.50m^3$
b. Paved area = $105.91 \times 2.9181 \times 0.7 = 216.34m^3$
c. 50% of a verticle wall = $\left(21.34 \times \frac{8.9}{2}\right) \times 2.9181 \times 0.7 = 193.98m^3$

Option I: Rooftop only = (546.27)/365 = 1.5 m^3/day i.e., 45m^3/mean month

Option II: All Possible sources [roof, paved area and 50% of a verticle wall] = (546.50+216.30+193.98)/365 = 2.62 m^3/day i.e., 78.62 m^3/mean month)

Step 2: Greywater generation

Greywater generation:

For the seven-storied building, 32 × 120 = 3840 liter/ day = 128 m^3/ mean month.

For the two-storied building, 9 × 120 = 1080 liter/ day = 36 m^3/ mean month.

Step 3: Designed water consumption

For the indoor in the seven-storied building, using rainwater and greywater for toilet flushing and laundry (as per Example problem 3.3):

The ultra-low flush toilets require only 6.0 liters/flush. Total demand for toilets flushing = 3600 litres/day = 960 litres/week

The typical front-loading washing machine requires 100 liters/load. The weekly rainwater usage for laundry = 1800 liters/week

Converting the weekly rainwater usage to the daily indoor rainwater demand (total) = $\frac{960+1800}{7}$ = 394.28 liters/day.

Thus, the desired pump flow rate of **175 lpm** ($\approx 0.0029\, m^3/s$) is recommended.

Step 4: Characterization of rainwater and greywater

Physical, chemical, and bacteriological characteristics of rainwater and greywater were determined and suggested for.

- For the seven-storied building: (a) sedimentation, filtration, and chlorination treatment for indoor usage, and (b) direct convey to a constructed wetland front of the building.
- For the two-storied building, directly conveyed to the bioswale retention.
- Regional standards for drinking water and wastewater reuse with direct and indirect contact are followed to identify treatment facilities.

Step 5: Design of treatment processes

The operations, processes, and systems needed to achieve the required level of treatment were selected according to the criteria: final use of the treated water, efficiency in removing contaminants, cost-effectiveness, and applicability on onsite/decentralized systems. Once selected, the treatment systems were designed to determine the investments needed for their implementation. Furthermore, the operation and maintenance costs of such methods should be estimated.

Step 6: Sizing of water storage tanks

Demand for the water from the integrated system would lead to the tank sizing. Different percentages of water coverage need to be tested considering consumers' choices.

5.11 Management Strategies

The general management strategies are:

- Contact or consumption should be restricted during greywater system maintenance. Hand gloves and hand washing are required while handling greywater filters and other parts.
- Strictly maintain greywater lines by labeling greywater plumbing, including garden hoses.
- Applying untreated greywater is not allowed onto the kitchen garden producing fruits and vegetables consumed raw (i.e., strawberries, lettuce, carrots).
- Ensure the greywater undergoes recycling is free from launder diapers or generated by patients with an infectious disease.
- Stored greywater should be used within 24 h before bacteria multiply. After 24 h, this turns to black water.
- Avoid overload the system, divert the excess greywater to the sewer.
- Divert greywater containing harmful chemicals to the sewer or septic system
- Avoid surface water contamination by disposing of greywater underground or into a mulch-filled basin.
- Avoid greywater application to saturated soils.

The following steps are recommended to ensure the greywater quality:

- Frequent inspection on pumps and valves to ensure that no bubbles have formed in the pump and to check that the valves are not blocked or damaged;
- Ensure uninterrupted electricity to the pumps;
- Frequently compare the inlet water to the outlet water;
- Constantly check greywater pipes to ensure the absence of solids collection and clogging in the lines;
- Remove sediments collected in the bottom of the greywater collection tank frequently (once a year);
- Observe system blocking indicators: a collection of water on the system's surface or unchangeable water level in the treated greywater collection tank, and track units' odor emission.

Following actions are recommended to improve the quality of treated greywater:

- If kitchen water is used, remove food waste from dishes and cooking pans before washing them. Use a sink strainer to prevent the food waste from mixing with water.
- Do not use chemicals in greywater sources like solid cleaning agents, paints, and pharmaceuticals.
- Do not clean children in greywater sources.

As for the health considerations in using greywater, the recommended actions are:

- Use treated greywater at its production site;
- Preferably, use treated greywater far from sites frequented by children and pedestrians;
- Place signs in areas where greywater is implemented to indicate the reuse of treated greywater in the area;
- Avoid storing greywater before or after treatment for more than 24 h to prevent bacterial growth and odor spread. Preferably, apply a ventilation pipe that allows the odour to escape, especially in collection tank where water may be stored;
- Do not use greywater in any practice that leads to direct contact with vegetables or plants that are eaten raw or cooked;
- Use drip irrigation methods and preclude irrigation by sprinklers;
- Avoid the use of treated greywater in an epidemic condition.
- Use proper greywater marker plates where applicable (Fig. 5.9).

Fig. 5.9 Typical reclaimed greywater marker plate

(a)

(b)

References

ADEQ (2002) Using gray water at home. Title 18, Chapter 9, Article 7. www.azsos.gov/public_services/title_18/18-09.pdf

Al-Jayyousi OR (2003) Greywater reuse: towards sustainable water management. Desalination 156(1):181–192. https://doi.org/10.1016/S0011-9164(03)00340-0

Ardern E, Lockett WT (1915) Experiments on the oxidation of sewage without the aid of filters, part III. J Soc Chem Ind 34:937.

Betriebs F, Darmstadt P (2005) Greywater recycling. fbr Dialog GmbH Havelstr. 7 A 64295 Darmstadt. www.fbr.deinfo@fbr.de.

BSI (2010) BS 8525-1:2010 greywater systems—code of practice. British Standards Institution.

Busser S et al. (2006) Characterisitcs and quantities of domestic wastewater in urban and peri-urban households in Hanoi. http://ir.library.osaka-u.ac.jp/dspace/bitstream/11094/13204/1/arfyjsps2006_395.pdf

Cameron RWF et al (2014) What's "cool" in the world of green façades? How plant choice influences the cooling properties of green walls. Build Environ 73:198–207. https://doi.org/10.1016/j.buildenv.2013.12.005

CMHC (2004) Water reuse standards and verification protocol. CHMC, Ottawa, 2004, Research Report 1611–06.

Crook J, Surampalli RY (1996) Water reclamation and reuse criteria in the U.S. Water Sci Technol 33(10):451–462. https://doi.org/10.1016/0273-1223(96)00448-9

Department of Health (2010) Code of practice for the reuse of greywater in Western Australia. https://www.gwig.org/the-greywater-guide-resource-pack

Eriksson E et al (2002) Characteristics of grey wastewater. Urban Water 4(1):85–104. https://doi.org/10.1016/S1462-0758(01)00064-4

Fane S, Reardon C (2013) Wastewater-reuse. https://www.yourhome.gov.au/water/wastewater-reuse. Accessed 13 August 2021.

Friedler E (2004) Quality of individual domestic greywater streams and its implication for on-site treatment and reuse possibilities. Environ Technol 25(9):997–1008, England. https://doi.org/10.1080/09593330.2004.9619393

Gregory JD et al (1996) Revision of the national reclaimed water guidelines. Desalination 106(1):263–268. https://doi.org/10.1016/S0011-9164(96)00117-8

Gross A et al (2007) Recycled vertical flow constructed wetland (RVFCW)—a novel method of recycling greywater for irrigation in small communities and households. Chemosphere 66(5):916–923. https://doi.org/10.1016/j.chemosphere.2006.06.006

Health Canada (2010) Canadian guidelines for domestic reclaimed water for use in toilet and urinal flushing. H128-1/10-602E.

Hernandez-Soriano MC, Jimenez-Lopez JC (2012) Effects of soil water content and organic matter addition on the speciation and bioavailability of heavy metals. Sci Total Environ 423:55–61. https://doi.org/10.1016/j.scitotenv.2012.02.033

Hoekstra AY et al (2011) The water footprint assessment manual: setting the global standard. Earthscan, London, UK.

IPCC (2008) In: Eggleston HS, Miwa K, Srivastava N, Tanabe K (eds.) 2006 IPCC guidelines for national greenhouse gas inventories—a primer, prepared by the national greenhouse gas inventories programme. IGES, Japan.

ISO (1995) ISO 6708:1995(en). Pipework components—definition and selection of DN (nominal size).

Jefferson B et al (2001) Treatment methods for greywater. In: Lens P, Zeeman G, Lettinga G (eds.) Decentralised sanitation and reuse: concepts, systems and implementation. IWA Publishing. London, UK, pp 334–353.

Jungels J et al (2013) Attitudes and aesthetic reactions toward green roofs in the Northeastern United States. Landsc Urban Plan 117:13–21. https://doi.org/10.1016/j.landurbplan.2013.04.013

Ledin A et al (2001) 'Aspects of groundwater recharge using grey wastewater. Chapter 18. In: Lens P, Zeeman G, Lettinga G (eds) Decentralised sanitation and reuse: Concepts, systems and implementation. IWA Publishing, pp 354–370.

Morel A, Diener S (2006) Greywater management in low and middle-income countries, review of different treatment systems for households or neighbourhoods. Swiss Federal Institute of Aquatic Science and Technology (Eawag). Dübendorf, Switzerland.

Nolde E (2000) Greywater reuse systems for toilet flushing in multi-storey buildings—over ten years experience in Berlin. Urban Water 1(4):275–284. https://doi.org/10.1016/S1462-0758(00)00023-6

NRMMC-EPHC (2006) National guidelines for water recycling: managing health and environmental risks. The Environment Protection and Heritage Council and Natural Resource Management Ministerial Council, Canberra, Australia.

NSW (2008) Interim NSW guidelines for management of private recycled water schemes. Department of Water and Energy, NSW.

Otterpohl R et al (1999) Source control in urban sanitation and waste management: ten systems with reuse of resources. Water Sci Technol 39(5):153–160. https://doi.org/10.2166/wst.1999.0234

Del Porto D, Steinfeld C (2000) The composting toilet system book (Concord, Massachusetts: Massachusetts The Center for Ecological Pollution Prevention (CEPP)), pp 235.

Queensland Government (2008) Greywater guidelines for Queensland councils. https://www.hpw.qld.gov.au/__data/assets/pdf_file/0008/4103/greywater-guidelines-councils.pdf

Samayamanthula DR et al (2019) Treatment and effective utilization of greywater. Appl Water Sci 9(4):90. https://doi.org/10.1007/s13201-019-0966-0

Schuetze T, Santiago-Fandiño V (2013) Quantitative assessment of water use efficiency in urban and domestic buildings. Water. https://doi.org/10.3390/w5031172

Shrestha RR et al (2001) Application of constructed wetlands for wastewater treatment in Nepal. Water Sci Technol 44(11–12):381–386. https://doi.org/10.2166/wst.2001.0855

State of California (2011) AB-849 water: use efficiency: graywater building standards.

Tajima A (2005) The behaviour of the pathogenic microbes in the treated wastewater reuse system and the establishment of the new technical standard for the reuse of treated wastewater. In: Proceedings of the IWA specialty conference on wastewater reclamation and reuse for sustainability. CD-rom.

TWDB (2005) The texas manual on rainwater harvesting, 3rd edn. Texas Water Development Board.

USEPA (2004) Guidelines for water reuse. USEPA, Washington, DC, 2004, Report EPA/625/ R-04/108

WHO (2006) Guidelines for the safe use of wastewater, excreta and greywater. Resource document. WHO. http://www.who.int/water_sanitation_health/wastewater/gsuww/en/index.html. Accessed 5 March 2007.

Wilderer PA (2004) Applying sustainable water management concepts in rural and urban areas: some thoughts about reasons, means and needs. Water Sci Technol A J Int Assoc Water Pollut Res 49(7):8–16, England.

Wright M (1996) Safe use of household greywater. Guide M-106. Cooperative Extension Service College of Agriculture and Home Economics. http://aces.nmsu.edu/pubs/_m/m-106.html

Chapter 6
Groundwater Recharge

6.1 Introduction

Groundwater recharge is a part of the hydrologic cycle; water moves downward through drainage or percolation from the water table to the saturated zone. In this process, water enters through an aquifer and encompasses water movement in the vadose zone. Groundwater recharges both naturally (i.e., hydrologic cycle) and artificially, as described in Chap. 2. In the managed aquifer recharge or artificial groundwater recharge, rainwater, or recycled water are considered to be routed into the subsurface. This chapter introduces the managed aquifer recharge, relevant technologies, and worldwide specific regulations and guidelines. For the managed aquifer recharge, both individual or combine the available form of water, i.e., rainwater, stormwater, and greywater, could be used. The managed aquifer recharge technologies for all of these three sources are described. So far, minimal water treatment is required while recharging with the rainwater. Recharge with stormwater and/grey water requires significant treatment. Regional guidelines, legislation, and regulations are formulated based on their desired water quality from the managed aquifer, potable or non-potable water. Working examples on the managed aquifer recharge feature the different water sources to recharge the aquifer. Thus, the possibilities and challenges of other technologies are incorporated in rainwater harvesting, stormwater, and greywater reuse to recharge aquifers are discussed in this chapter.

6.2 Managed Aquifer Recharge

Flow underneath the land surface proceeds through the infiltration process within the hydrologic cycle. The water supply for prolonged periods maintains entirely saturated soil with water. If intermittent water supply persists, the absence of recharge occurs before the first infiltration or between two subsequent infiltrations. The water movement in soil within two infiltration occasions is known as redistribution.

© The Author(s), under exclusive license to Springer Nature Switzerland AG 2022

A. Akter, *Rainwater Harvesting—Building a Water Smart City*, Springer Water,
https://doi.org/10.1007/978-3-030-94643-2_6

Recharge also occurs in the absence of a hydraulic connection between the ground surface and the underlying aquifer. The artificial recharge generates either a piezometric effect or volumetric effect. The *piezometric effect* results due to the increased piezometric surface.

The *storage coefficient* is the ratio of the aquifer's transmissivity and the replenishment coefficient. The piezometric effect is influenced by the aquifer's capillary forces, water temperature, and air voids. The *volumetric effect* is linked to the aquifer's specific yield, replenishment coefficient, transmissivity, geologic and hydraulic properties. The bulk recharge water moves through spreading, or a sliding resulted speed-related to groundwater flow.

6.3 History

'Artificial recharge', 'enhanced recharge', 'water banking' or 'management of aquifer recharge and subsurface storage' has a glorious history. In this connection, "Managed Aquifer Recharge" (MAR) was first named in 2005 (Gale and Dillon 2005). The MAR development was achieved into four stages (Zhang et al. 2020), and these are (Fig 6.1):

The first stage (221 BC–1850 AD) was initiated at a minor development and limited application using infiltration channel focused on increasing groundwater level for agriculture irrigation and improving groundwater quality. The idea of MAR was invented in China in 221 BC, starting with the dug well concept during the Warring States of China (475 BC–221 BC) (Wang et al. 2014). The Qin and Han Dynasties (221 BC–220 AD) used underground drains, channels, and waterlogged areas in China. After that, "Amunas," an infiltration channel practiced by the Wari (a pre-Inca civilization) from 500 to 1000 AD in Peru (Gammie and Bievre 2015). Then, Careo (infiltration channels) was invented in Spain in the eleventh century (Memola 2014).

The second stage (1850–1950) aimed to meet water demand by industries and restore the surface quality during the progressive industrialization in the nineteenth century. Riverbank filtration was applied in the UK, the Netherlands, and Finland in 1870 (Stuyfzand 2015). Around 1900, infiltration ponds for storm runoff were performed in California, US, and widely practiced in the 1930s (Weeks 2002).

In the third stage (1950–1990), i.e., after World War II, the water quantity and quality during supply were prioritized due to post-war refurbishment and urbanization. The Aquifer Storage and Recovery (ASR) were conceptualized and implemented in due course. In 1968, the long-term ASR well field was executed to minimize salinity intrusion in New Jersey, US (David and Pyne 2005). Then in Europe, large-scale MAR projects for water supply in the Netherlands were started in the 1950s. In the 1960s, large-scale MAR projects were established in Australia, Finland, and China (Charlesworth et al. 2002; Wang et al. 2014).

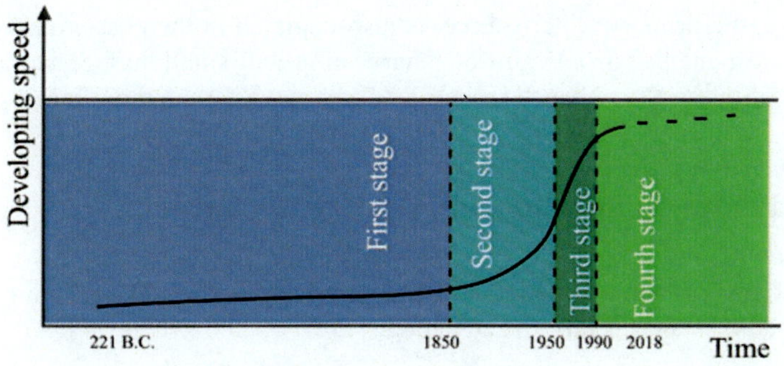

Fig. 6.1 Historical development of MAR (Zhang et al. 2020)

Fourth stage (1990 to the present) MAR projects have been applied in developed and developing countries to contribute to water reserves considering climate variability, increased population growth, and ecological and environmental requirements. In this stage, the developed countries (Europe, US, UK, and Australia) were advancing the technical specifications compared to the developing countries (i.e., India, South Africa, and China) efforts on design and development macro-strategies formulation. In 2007, the Indian government allocated funds for large-scale dug-well to restore water in hard-rock aquifers that experienced over-exploitation.

6.4 Need for Managed Aquifer Recharge

Natural replenishment of groundwater reservoirs progresses through infiltration (as stated in Chap. 2). Due to the slow process, this often fails to balance the extreme and persistent exploitation of groundwater. Thus, water levels drop and decline groundwater resources. MAR enhances the surface water movement naturally into groundwater reservoirs through suitable construction approaches. These approaches correlate and incorporate the water sources to the groundwater reservoir but are strongly influenced by the hydrogeological properties of the concerned area.

- Suppose the climatic conditions are unable to create surface storage. In that case, MAR techniques are adopted for diverting surface runoff to the groundwater reservoirs within the quickest possible time.
- In arid regions, the mean annual PET is higher than the available precipitation. Planning for the annual water resource should be prioritized rain conservation through surface or sub-surface reservoirs.
- In hilly areas with high rainfall, water scarcity is experienced post-monsoon due to the majority of the rain turns to surface runoff.

- Springs, the primary water source, is also depleted in the post-monsoon. These areas are benefited from rainwater harvesting and small surface storage at the points of interest to enhance recharge during and after the rainy season.

6.5 Advantages of Managed Aquifer Recharge

The MAR ensures sustainable groundwater provisions to meet the growing demands in the urbanized areas. Thus, the advantages are:

- Reduce water logging or inundation due to urban stormwater;
- Provide space for subsurface storage;
- Negligible Evaporation losses;
- Enhanced infiltration by providing the permeable media;
- Higher biological purity;
- Social impacts are absent as there is no requirement for population displacement nor intake of agricultural land, etc.
- Minimum temperature variations;
- The MAR is a soil conservation approach, also offers adequate soil moisture in summer;
- The provision of underground water storage is relatively safe against any natural and artificial disasters;
- The MAR ensures natural allocation among recharge and withdrawal points;
- Energy conservations in suction and delivery head reduction utilizing increased groundwater levels.

6.6 Types of Managed Aquifer Recharge

Artificial aquifer recharge has been practicing for centuries. Initially, only the fresh surface water has been replenished; after that, rainwater or greywater continues to reserve groundwater for further intensive use/abstraction.

- Infiltration techniques
  - Soakaway/trench
  - Swale
  - Infiltration gallery
  - Permeable pavements
  - Infiltration basin
  - Ditch and furrow system
- Direct recharge techniques
  - Open well/borehole
  - Recharge shaft

  - Injection bore
- Channel modification techniques
  - Check dams
  - Recharge dams
  - Streambed modification
- Catchment management
  - Contour bunding
  - Gully plugging
  - Afforestation
  - Controlled grazing
- Indirect recharge techniques
  - Induced recharge
  - Over irrigation
  - Leaking water and wastewater pipe networks
  - Sewage disposal by septic tank

Infiltration techniques intrude water into the vadose (unsaturated) zone; then, water percolation towards the water table across the soil and rock, providing a natural contamination remedy. These techniques are generally applicable and straightforward but require suitable ground and hydrogeological requirements.

6.6.1 Rainwater Harvesting for Managed Aquifer Recharge

Based on the hydrological and hydrogeological circumstances, the commonly practiced techniques are:

Recharge pits

Recharge pits follow the concept of the '*soakaway*,' i.e., the traditional disposal system of stormwater runoff from buildings and other hardstand remote areas away from a proper drain or public sewer. These are suitable for shallow aquifers and sandy/permeable strata recharge. Thus, groundwater recharge by constructing over-ground or underground recharge pit would be appropriate if the areas are of exposed (i) silty material overlies the aquifers, (ii) aquifers containing shallow depth (5–15 m), (iii) peat soils, or (iv) soil containing organic matter.

Surface runoff from the city is diverted into the recharge pit. These pits are comprised of boulders (backfilled), gravels (in between), and coarse sand (at the top). Silt-containing runoff passed through the rough sand layer and removal of silt deposition from the top layer. Broken bricks or cobbles are used as filling materials to serve lower surface runoff. For better performance, pre-filtration, i.e., screening and

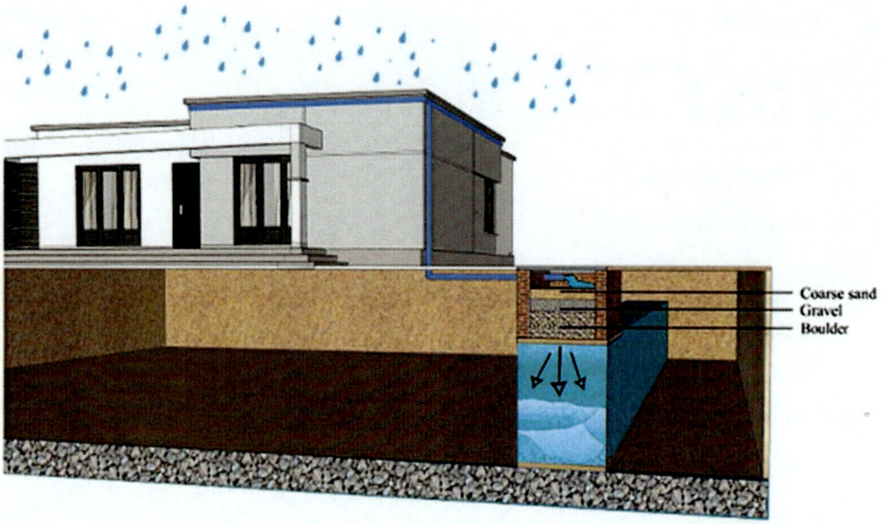

Fig. 6.2 Recharge pits

first flush, as described for the rainwater harvesting, should be conducted following Sect. 3.6 in Chap. 3.

Recharge pits are widely available around the world, and there are specific guidelines and design standards. For instance, in Auckland, design guidance for all soakage systems (Auckland Council 2013) is available in India for a 100 m^2 rooftop; the suggested recharge pit configurations as illustrated in Fig. 6.2.

Infiltration that occurs in the recharge pit depends upon:

- The groundwater table—the lower, the better;
- Land use/Landcover (LULC); and
- Underlying soil properties—preferable for a high percolation rate.

Infiltration testing for the recharge pit follows the following procedures (Bettess 1996):

i. Trail pit excavation with proper dimension;
ii. Records of the wetted perimeter for the half-full pit;
iii. Pit fills up with the invert level of the inflow pipe;
iv. Records of water level/depth at frequent time intervals;
v. Total three tests are recommended, preferably on the same day. The lowest infiltration rate uses to design the soakaway:

Thus, the acquired infiltration rate, $f\left(\frac{mm}{h}\right)$:

$$f = \frac{V_{75-25}}{a_p \times t_{75-25}}. \tag{6.1}$$

Here, between 75 and 25% of the depth to the maximum level.
V_{75-25} = Storage volume (m^3)
t_{75-25} = Time elapses for the pit to empty (h).
Theoretically available water volume (m^3), V_t
= Maximum volume of water available for recharge (m^3)

$$= A \times P_{avg} \times C \tag{6.2}$$

The total volume of water would be recharged through ‘n’ numbers of pits,

$$V_{Rp} = \sum \left(V_{Rp_1} + V_{Rp_2} + V_{Rp_3} + \cdots + V_{Rp_n} \right) \tag{6.3}$$

The volume of water that each pit might recharge,

$$V_{RP_1} = C \times a_{p1} \times \overline{P} \tag{6.4}$$

$$\text{Depth of recharge (soak) pits, } h = \frac{1}{a} \times \left(A \times C \times \overline{P} - K \times a \right) \tag{6.5}$$

P_{avg} = Mean annual precipitation (m) from the 20 years monthly data
$\overline{P}$ = Mean annual rainfall (m)
A = Potential area for recharge (m^2)
a = Area of pit (m^2)
a_p = catchment area served by each pit (m^2)
C = Runoff coefficient
K = Hydraulic conductivity, values are available in Table B.1, Appendix B.
The total water volume naturally recharged by ‘n’ numbers of pits,

$$V_{RfN} = \sum \left(V_{Rf_1} + V_{Rf_2} + V_{Rf_3} + \cdots + V_{Rf_n} \right) \tag{6.6}$$

The volume of water that might be recharged naturally by each pit,

$$V_{Rf_1} = C \times a_{base,p1} \times f \tag{6.7}$$

Here,
f = Infiltration rate of each pit (mm/hour) [values are available in Table 2.4, Chap. 2].
$a_{base,p1}$ = The average area of the base of the height of the pit (m^2).
The volume of water fail to be recharged or wasted through pits,

$$V_w = V_{RP} - V_{Rfn} \tag{6.8}$$

The pit capacity depends on the catchment area, rainfall intensity, and soil recharge rate. Generally, the pit’s dimensions are 1–2 m in width and 2–3 m deep, based on the available thickness of the permeable strata. These pits are appropriate for shallow aquifers recharging.

Recharge Trenches

With similar mechanisms of recharge pit but aimed to serve more extensive area recharge trenches are used. Recharge trenches are known as 'Infiltration Trench' or 'Contour Trenching'or 'Soakaway' or 'Trench Soakaways.' With the sufficient infiltration capacities of the surrounding soil, the trench is designed to absorb the runoff from a subsequent storm. These are applied to the preamble strata at shallow depths by providing three layers, i.e., boulders (backfilled), gravel (in between), and coarse sand (at the top), thus capture silt contents while surface runoff passed through (Fig. 6.3). Although the maintenance requirements are generally low, clogging the system could slow down recharges; thus, there is a greater risk of failure. For better performance, trenches should need cleaning and desilting regularly.

In semi-arid regions, experience from India showed that shallow intermittent trenches dug perpendicular to the land slope of small-scale earthen embankment or vegetation can infiltrate up to 50% of peak rainfall. The benefits of wetlands are achieved in the soakaway skills by adopting recharge trenches or pits.

Open Well/Borehole

Areas with dry shallow aquifers require tapping the existing wells into the deeper aquifer, the current abandoned open well, or borehole uses for shallow or deep aquifers recharging. Historically, a *dug well* was excavated below the groundwater table, and, to prevent collapse, stones, brick, tile, or other materials are used to line (cased) the well (Fig. 6.4). Then, optionally a cover of wood, stone, or concrete uses to protect the well. The harvested rainwater should be made to pass through pre-filtration before discharging it into the well by a T-joint and then crosses multiple filter

Fig. 6.3 Recharge trenches

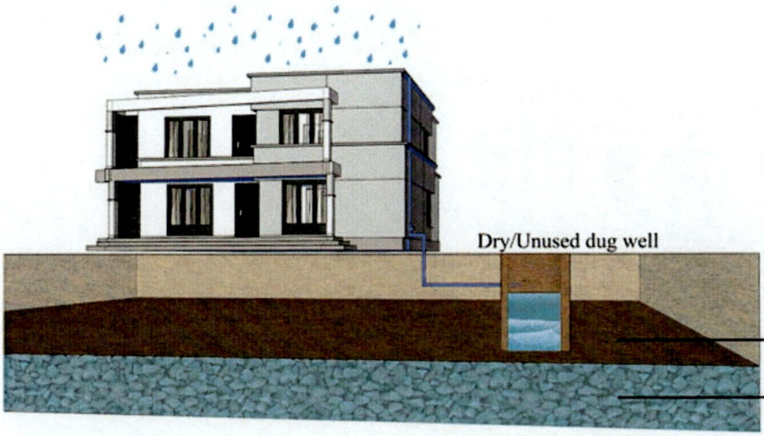

Fig. 6.4 Dug wells

chambers. The filter has reducers on both sides. The selection of filter material and plumbing requirements are based on the catchment size. There is a guideline by US EPA for the new dug well construction to minimize the likelihood of contamination.

For larger catchment areas, downpipes collect and covey surface runoff to the over groundwater chambers. These chambers are interconnected and forwarded the collected water to the filter pit through pipes. A connecting pipeline with a recharge well is provided at the pit bottom for recharging filtered water through the well. Figure 6.5 shows single-home *domestic water well,* either hand-pumped or mechanically pumped, is installed to access filtered water and monitor filtered water quality.

Recharge Wells

Impervious surface soil usually generates significant surface runoff within a short duration of heavy rainfall, and wells are constructed to recharge the groundwater storage through trench/pits. These are suitable for the permeable horizon within 3 m of ground level and applied for deeper aquifers. They have required drilling a borehole depth of 3 m above the groundwater table and a suitable bore diameter based on the receiving strata porosity. Construction of the recharge well maintains 3–5 m below the water level. Slotted pipes and gravel packs are installed within the borehole to enable permeability (Fig. 6.6). Recharge well is suggested to place at least 10–15 m away from the buildings. If the gravity force fails to recharge the aquifer, an *injection bore/injection well* is usually constructed to pass directly into the deep aquifer through the screens provided in the well due to the pressure. *Direct recharge* enters water into the phreatic or saturated zone of the aquifer. These techniques require less land surface area and cost-effective capital cost. For the *Safe Drinking Water* Act, the US EPA has set the minimum standards to avoid threats for injection wells; these are sediment, nutrients, metals, salts, microorganisms, fertilizers, and pesticides.

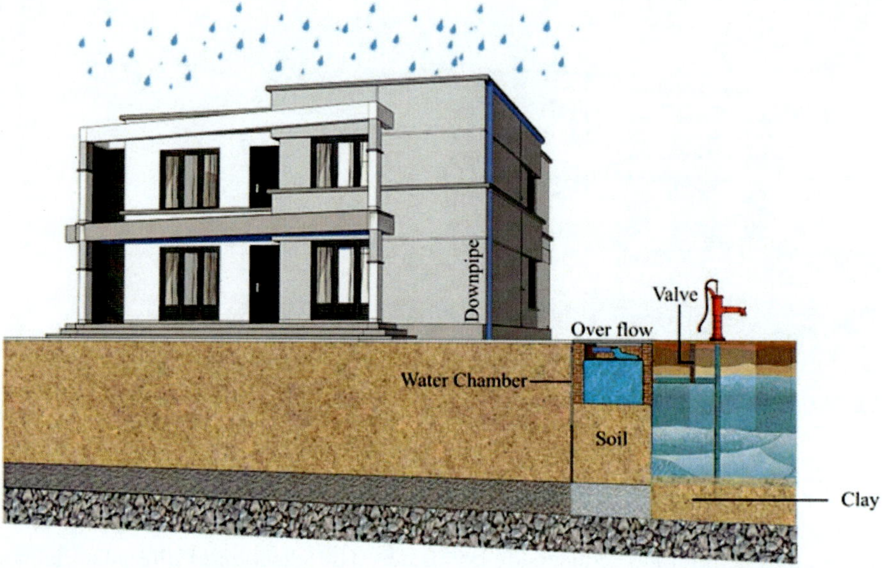

Fig. 6.5 Hand pumps

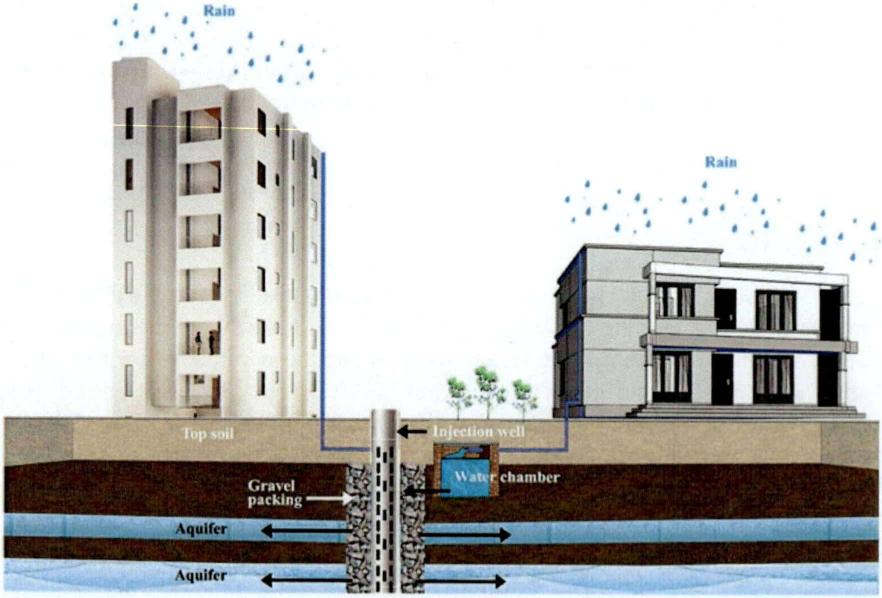

Fig. 6.6 Recharge well

The volume of available water would be recharged through each well,

$$V_{Rw_1} = a_{w1} \times \overline{P} \times C \tag{6.9}$$

where,

a_{w1} = Catchment area served by each well (m^2)

The total volume of available water would be recharged through 'n' numbers of wells:

$$V_{Rw} = \sum \left(V_{Rw_1} + V_{Rw_2} + V_{Rw_3} + \cdots + V_{Rw_n} \right) \tag{6.10}$$

Here, the available maximum volume of water that each well could be naturally recharged:

$$V_{N(w1)} = V_{AvailableN(w1)} = V_{max} - V_{min} = \left(h_{max,w1} - h_{min,w1} \right) \times A_{base(w1)}$$

Thus, naturally recharged volume through 'n' number of wells:

$$\begin{aligned} V_{N(wn)} &= \sum V_{N(w1)} + V_{N(w2)} + V_{N(w3)} + \cdots + V_{N(wn)} \\ &= \sum (h_{max} - h_{min}) \times A_{base(w)} \end{aligned}$$

Here,

$h_{max,n}$ = Maximum height of water level in the *n*th well per year

$h_{min,n}$ = Minimum height of water level in the *n*th well per year

$V_{N(wn)}$ = Maximum volume of water available in the expected recharge.

The available average water volume for recharge in each well, $v_{avg(w1)N}$. This recharge is practically possible available water volume for recharge in each well.

$$v_{avg(w1)N} = h_{avg} \times A_{base(w1)} \tag{6.11}$$

h_{avg} = Average height of water level.

The average volume of water available for recharge due to the 'n' number of wells

$$V_{(avg)N} = v_{avg(w1)N} + v_{avg(w2)N} + \cdots + v_{avg(wn)N} \tag{6.12}$$

The available water volume for recharge through wells (n') that are not considered, V'

$$V' = n' \times h'_{avg} \times A'_{base(avg)} \tag{6.13}$$

Here,

h'_{avg} = Average height of water level fluctuation in each well that is not considered

$A'_{base(avg)}$ = Average area of the base of the well that is not considered.

Thus, the available total water volume would be recharged during normal conditions:

$$V_T = V_{Rw} - V' \tag{6.14}$$

The total volume in recharge condition after rainwater harvesting:

$$V_{totalR(w)} = v_{R(w1)} + v_{R(w2)} + \cdots + v_{R(wn)} \tag{6.15}$$

Here,

$v_{R(wn)}$, the available water volume during the rainy season for recharge per well. It is the practically possible volume of water that could be trapped for recharging during the rainy season.

$$v_{R(w1)} = V_{\max(R,w1)} - V_{\min(R,w2)} = \left(h_{\max,\text{w1}} - h_{min,w2}\right) \times A_{base(w1)}$$

Here, $h_{\max}$ and h_{min} are the maximum and minimum water levels respectively in storms.

Recharge Shafts

Recharge shafts are constructed to restore poorly permeable strata-covered unconfined aquifer (Fig. 6.7). The diameter and depths of recharge shafts are selected based on runoff availability. These shafts contain boulders, gravels, coarse sand, and sand fills. *Vertical recharge shaft structures* and the artificial lithology layers are constructed to convey low surface runoff through hard rock and older alluvial plains. Aquifers are available at greater depth (viz. 20 or 30 m); a recharge well would be

Fig. 6.7 Recharge shafts

built within the shaft for recharging the open water to the deeper aquifer. A filter media is provided at the bottom to avoid congestion of the recharge well. The *shaft with a recharge well* is suitable for a catchment area of 1500 m^2 and more. On the other hand, a *lateral trench* or *shaft with a bore well* recharges deeper aquifers and is ideal for a catchment area over 5000 m^2.

6.6.2 Managed Aquifer Recharge with Stormwater

Stormwater drainage wells enhance subsurface infiltration capacity to ensure stormwater surface runoff. MAR is categorized broadly into three depending on the penetration depth of the recharge structure: surface infiltration devices, vadose zone infiltration devices, and injection wells. The detailed design on the bioretention swales, bioretention basin, and the permeable pavement has been described in Sects. 4.5.3, 4.5.4, and 4.5.5, respectively (Chap. 4). *Permeable paving* is relatively expensive, and the permeability of the paving-supported infiltration capacity needs to handle the rainfall intensity. US EPA underground injection control regulations (USEPA 2014) describe the stormwater drainage well requirements. Typically, stormwater drainage wells include dry wells, bored wells, and infiltration galleries.

Drywells

Drywells are *vadose zone* infiltration wells constructed within a low permeable soil layer to dispose of stormwater runoff. The collected surface runoff is subjected to pre-treatment through a grass swale before storing in a sedimentation tank. Then, recharge the aquifer to augment groundwater resources through a stilling pipe. These wells are generally constructed in areas having deep water tables. Perforations in dry wells casing infiltrate water through the unsaturated zone towards the unconfined aquifer (Fig. 6.8). Required gravel fill/packing is provided to minimize the clogging of the pipe. Drywell structures are widely practiced around the world.

Infiltration galleries

The working principle of an infiltration gallery is similar to a trench soakaway, which temporarily reserves water within the pipes and adjacent gravels during infiltration progresses through the soil/strata. A comprehensive horizontal perforated pipes cover 10% of the surface area surrounded by a gravel pack linked to convey vertical inflows from the ground surface. Geotextiles or impermeable plastic sheets protect the top and side of the galleries from unwanted direct percolation. These protective materials also restrict the entrance of soil and other fine particles within the gallery. Infiltration galleries require minimal land surface footprint; thus, they are responsive to horizontal groundwater abstraction from aquifers (Fig. 6.9). The duration is approximately 75 days from the water entry through the inflow pipe. Maintenance is similar to soakaways, and inspection maintenance holes are mandatory to update the systems. Examples are available in parks in New Delhi and a 30 km irrigation canal in Gujurat.

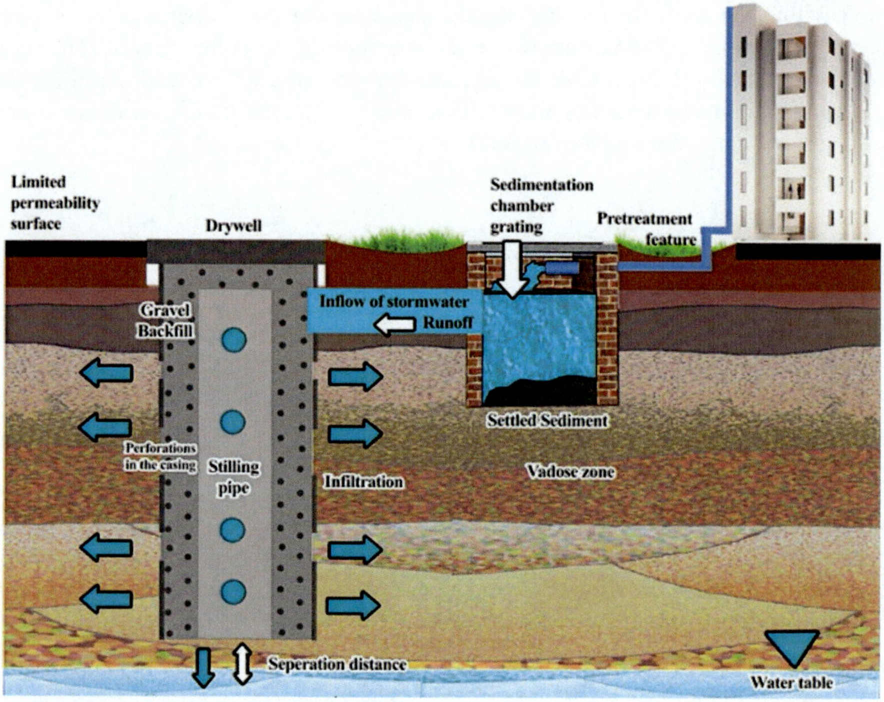

Fig. 6.8 The design of a typical drywall

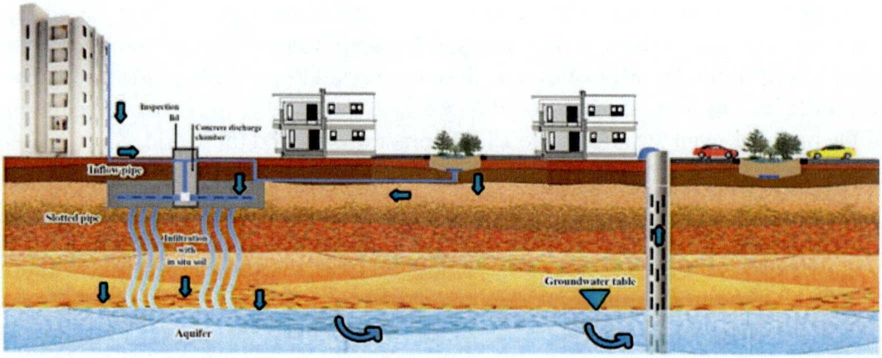

Fig. 6.9 Infiltration gallery

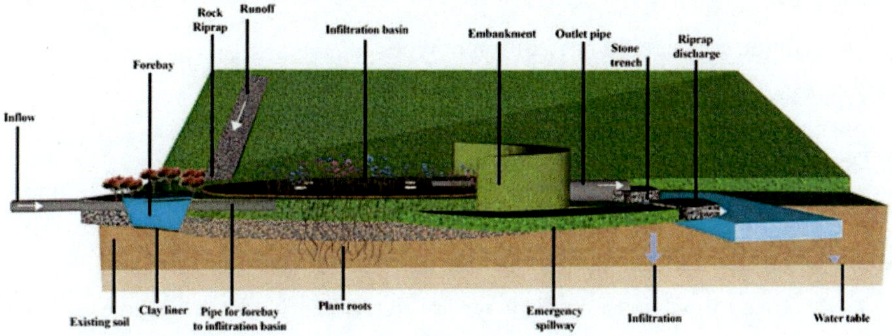

Fig. 6.10 Infiltration basins

6.6.3 Managed Aquifer Recharge with Stormwater and Greywater

Infiltration basins are usually constructed to reduce urban flooding through surface runoff and greywater management and contribute to artificial aquifers' recharges. Recharge wells and shafts are built to augment infiltration and restore lower permeability superficial formations of soil structures near the ground surface. Due to their desilting design, they are often prone to clogs, silts, deposition of suspended solids. Along with the underlying soil/strata characteristics and evaporation losses, their efficiency varies as recharge structures.

Inflow from the storm sewer and surface runoff store combine to a forebay via rock riprap passage. Then, a connected pipe from the forebay would convey the water towards the infiltration basin (Fig. 6.10). Other components include: rock riprap, vegetation, and an outlet.

Maintenance of the infiltration should reduce the clogging layer through scraping or digging depending on the size in the system will restore infiltration rates. Occasional storm flows with no inter-storm flow within the wadis/canyons in the Middle East faces difficulties in handling higher sediment loads within the storm flows. Then, infiltration basins have been excavated below the gradient of check dams to prevent immediate clogging. The check dam constructs to retain the storm flow, decline velocity, and promote sediment settlement within the floodwaters.

Infiltration basins differ from the bioretention basins, and the key issues are:

Infiltration basin (or trench, underground, dry well)

- In a watershed typically positioned downgradient of other water supply practices;
- Centralized or decentralized scale treatment persists;
- Designed for the rare significant storm that exceeds the capacity of upgradient methods;
- Maximum drainage area covers 0.202 sq.km with the maximum ponding depth of 1.22 m; and

- The growing medium is the native soil.

Bioretention basin could have different types

- Typically existed in the watershed;
- Onsite treatment control scale;
- Designed for the minor storms (water quality events);
- Maximum drainage area 20,234 m^2. with the maximum ponding depth of 0.305–0.457 m.
- The growing medium is the engineered growing medium.

As discussed in Chap. 5, *soakaways* are traditionally used in remote places to dispose of greywater from buildings and other paved areas. With a similar working principle, soakaway is designed to store immediate storm runoff or rainwater. Allowance of adequate infiltration through the adjacent soil confirms the soakaway to handle the subsequent storm. Rainfall, regional LULC maps, and geological maps are compiled to identify suitable soakaways sites for handle overflows from the storage tanks. Areas of highly permeable underlying sediments, and nearshore coral sands, are not recommended sites. Estimation of infiltration capacity would help to soakaways design. Soakaways are in a square or circular shape, filled with rubble, rock, or gravel. Trench-type soakaways are practicing for the larger areas. *Auckland Council technical report TR2013/040* (Auckland Council 2013) describes design procedure of soakage systems.

Aquifer Storage and Recovery (ASR) includes injecting storm/wastewater to the confined or unconfined aquifer in the wet season for storage and recovering from the same well in the dry season (Fig. 6.11). Examples are water wells in Florida and South Australia; dug wells in India based on ASR ensure perennial water table. These recharge schemes often face clogging inside the borehole screens and gravel

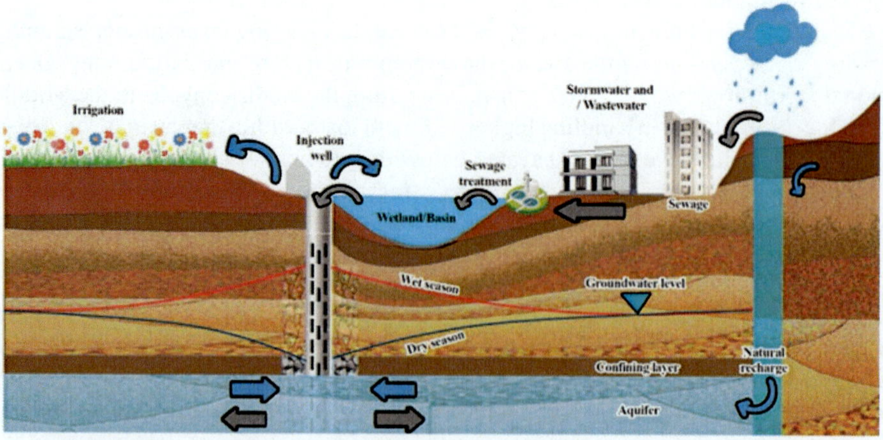

Fig. 6.11 Aquifer storage and recovery (USEPA 2016). Redline: storm/wastewater to the aquifer in the wet season. Blueline: recovery from an aquifer in the dry season

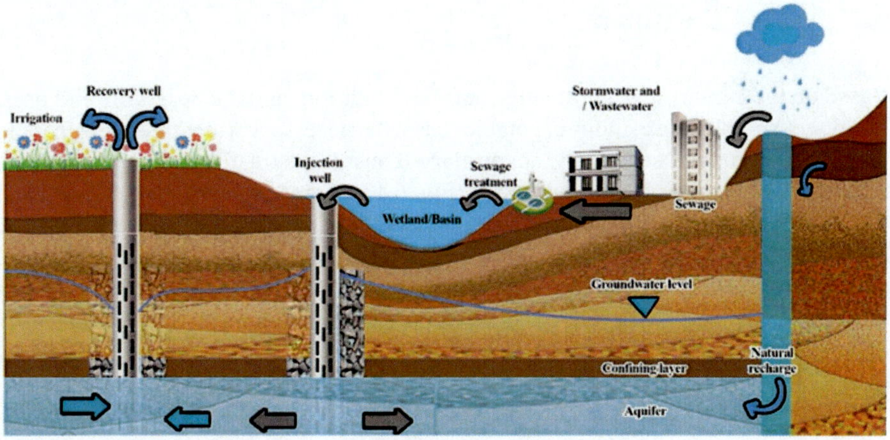

Fig. 6.12 Aquifer storage, transport, and recovery (ASTR)

pack due to bacterial growth, chemical precipitation, and silt deposition. Recharge water in the infiltration technique needs the necessary pre-treatment to maintain the chemical/biological composition that undergoes the natural treatment. Thus, reduce aquifer contamination as well as operation maintenance costs.

Aquifer storage, transport, and recovery (ASTR) also include injecting water into a well for storage and recovery from a separate well through an additional water treatment (Fig. 6.12).

If the injection recharge rate is 'a' liter per second, the thumb rule for India is 'a' liter per second, the expected number of days of recharge is b-days (CGWB 2007). Then, the quantity of recharge is:

$$Recharge(Million\ cubic\ meter) = \frac{a \times 86.4 \times b}{10^6} \tag{6.16}$$

Assume the injection rate of 5 liter/s, then 0.15 Million m^3 of water artificially recharge annually by each injection well.

Soil Aquifer Treatment (SAT) is also a MAR option intermittently infiltrated the treated sewage effluent through infiltration ponds to accelerate nutrient and pathogen elimination. The effluent travels across the unsaturated zone and is retrieved by wells. Examples are Alice Springs (Northern Territory, Australia), Arizona, and California in the US. These are addressed in the Australian Guidelines for MAR (Page et al. 2011), then extensive monitoring on water quality is required.

6.7 Water Treatment

Based on the capturing water quality, aquifer condition, and the deserving end uses, i.e., direct potable reuse, indirect potable reuse, the necessary water treatment in three phases, i.e., primary, secondary, and tertiary. Thus, these treatments comprise physical, chemical, and biological processes (Fig. 6.13). Typically existed unplanned and uncontrolled urbanization often treated incidental methods (Fig. 6.13a). On the other hand, economic involvements aimed at lowering groundwater pollution risk at source, the adapted planned treatment process includes infiltration basin/recharge lagoons (Fig. 6.13b). Recharge lagoons are custom-built impoundments in the recharge areas of aquifers, usually subjected to both secondary and tertiary treatment processes.

a

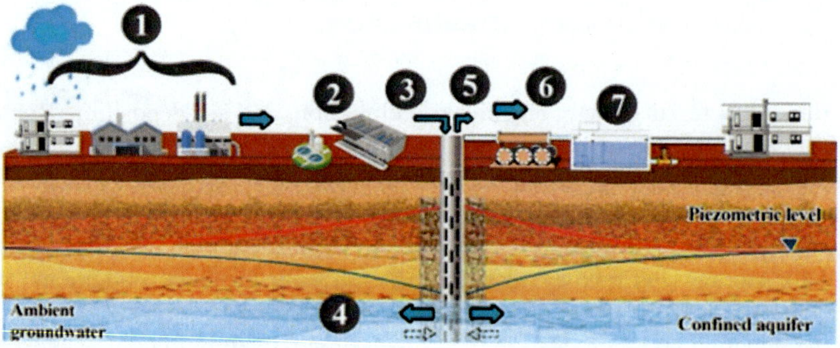

1: Capture zone, 2: Pre-treatment, 3: Recharge, 4: Subsurface storage, 5: Recovery, 6: Post-treatment, and 7: Storage for end-use.

b

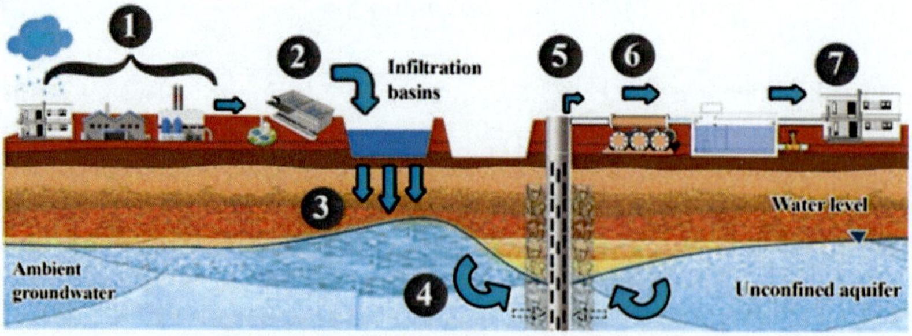

1: Capture zone, 2: Pre-treatment, 3: Recharge, 4: Subsurface storage, 5: Recovery, 6: Post-treatment, and 7: Storage for end-use.

Fig. 6.13 **a** Confined aquifer (ASCE 2020), **b** unconfined aquifer (ASCE 2020)

Wastewaters recharge lagoons can be used, and improvements in bacterial and chemical quality are possible in the recharge process (Coliforms, NO_3, etc.). The treatment process in confined and unconfined aquifers comprises seven steps: 1: Capture zone, 2: Pre-treatment, 3: Recharge, 4: Subsurface storage, 5: Recovery, 6: Post-treatment, and 7: Storage for end-use. Among them, post-treatment is only suggested for drinking water.

Treatments cover the solid setting, photolysis eliminates organic and pathogens, enhanced iron filtration removes phosphorous and metals, biofilter excludes trace organics, and denitrification while passing through geo-media before groundwater recharge.

- *Desalination* should be commenced either pre-treatment or post-treatment.
- *Organic matter* is removed through biodegradation, microbial assimilation, filtration tanks, sorption, or precipitation.
- *Nitrogen* can be removed by SAT merges dry and wet cycles to offer alternate aerobic and anaerobic conditions. Retrieval of *organic carbon* by Granular Activated Carbon (GAC) and membrane filtration. Subsurface organic carbon and nitrogen eliminates through *redox processes*. The *redox* (or oxidation–reduction) *process* is a chemical reaction that transfers electrons between two species. Validation should be supported by declining concentrations and physiochemical requirements
- *Nutrients* might be eliminated by in-line filtration using source-water delivery infrastructure (for particulate organic carbon).
- *Pathogen levels* so far are not reported in association with stormwater reuse. Although to further reduce pathogen levels, most schemes in the developed countries incorporate some form of treatment (e.g., constructed wetlands or biofiltration) while implementing managed aquifer recharge.
- *Phosphorus* removal technology should decline concentrations supported by mineralogy (iron, aluminium oxides) or mineral diffusion computations.

Example Problem 6.1 With a surface water-based improved water supply system, the seven-storied building stated in the *Example problem 3.3* decided to be a part of groundwater augmentation. The stakeholders are seeking options to introduce proper MAR technique with the rooftop rainwater harvesting.

Solution

As per *Example problem 3.3*, the available rainwater from the roof, open paved area, and 50% of a verticle wall are 546.50 m^3, 216.30 m^3 and 677.83 m^3 respectively. Considering most of the harvesting would be during the six months, i.e., May to October. To reduce the treatment costs, the arrangement would be different, i.e., recharging with:

Option 1: no treatment for the harvested rooftop rainwater. The desired recharge structure volume is:
$\frac{546.50\,\text{m}^3}{180}$ i.e. 3.04 m^3/day. To meet this capacity, a suggested recharge well size is of (2.55 m × 1.1 m × 1.1 m) i.e., 3.085 m^3.

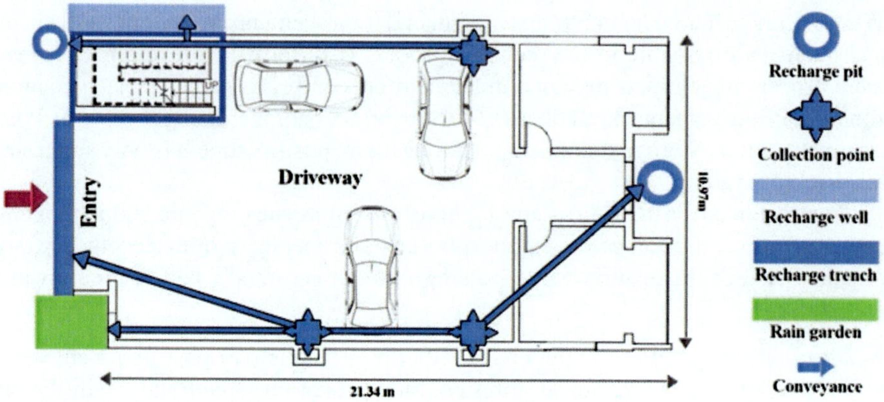

Fig. 6.14 MAR options on plan

Option 2: surface runoff from the paved area and the vertical wall are subjected to the necessary treatment. The desired recharge structure capacity is: $\frac{(216.30+677.83)\ \text{m}^3}{180}$ i.e., 4.97 m^3/day. This can be ensured that the size of the *two recharge pits* is (0.5 m × 0.5 m × 0.5 m), i.e., 0.125 m^3, and the *recharge trench* of (4 m × 1.1 m × 1.1 m), i.e., 4.84 m^3.

Option 3: If only artificial recharge is considered, stormwater runoff can be reduced/managed with a green roof in the roof area and porous pavements at both the ground floor front garden. The two options mentioned above would be ignored. The proposed *rain garden* should be 8.04 m^2. Thus, the front garden of 8.1 m^2 area can easily be converted into a rain garden with 1 m depth.

Thus, a schematic of the proposed MAR has shown in Fig. 6.14.

Nearly 0.15 MCM of water with the injection rate of 5 liter/s could be artificially recharged through each injection well in a year.

6.8 Aquifer Recharge Specific Regulations and Guidelines

The available MAR guidelines provide principles and a framework for the safe implementation of recycled water schemes and support the transition from after-treatment testing to a more integrated approach. Examples from different countries are:

Australia: The Australian Government has issued National Water Quality Management Strategy on water recycling guidelines on specialized requirements for drinking water augmentation through MAR schemes and environmental risk management.

Europe: *Urban Wastewater Treatment Directive* (91/271/EEC) recommends the wastewater treatment and minimum quality before recharge *Water Framework Directive* (2000/60/EC) defines the water quality specifications for reclaimed water at the point of recharge or withdrawal. Groundwater Directive (2006/118/EC) restricts hazardous substances and degradation and attenuation processes during soil passage.

Spanish Government has established legal regulations for the reuse of treated wastewater. The regulation (REAL DECRETO 1620/2007) specifies wide reclaimed water uses and water quality criteria for different applications.

US: The regulations issued by the California Department of Public Health is concerning wastewater reuse. Reclaimed water used for groundwater recharge should undergo prescribed treatment, sampling procedures, catchment management, and meet the water quality specifications. For groundwater recharge and reuse projects augmenting domestic water supply aquifers, comprehensive monitoring and control of the recycled water are specified. An arbitrary minimum residence time of 60 days has been suggested. However, this duration depends on the site-specific conditions.

Asia: Bangladesh's National Sustainable Development Goal 6.1 Action Plan has included MAR.

Thus, few countries prefer their regulations have legal status (i.e., Spanish and Californian). Rules are also practicing as non-binding and advisory (i.e., Australian). Wastewater recharges always rely on proper treatment and routine monitoring.

Example Problem 6.2 A two-storied building was constructed within the same plot as the seven-storied building stated in Example problem 3.3. Now, the entire compound is paved (Fig. 6.15). The owner attempted to compare the available MAR technique on possible recharge quantity?

Solution

Total available rechargeable water volumes are:

Daily available water

Source I: Harvestable rainwater from the building

Seven storied buildings (roof + 50% of a vertical wall) = (546.50 + 677.83) = 1223.83 m^3 [described in the Example problem 6.1].

Two storied building (roof + 50% of a vertical wall) = (546.50 + 193.98) = 740.48 m^3 [described in the example problem 5.2].

Total = 1223.83 + 740.48 = 1964.31 m^3.

The total amount could be recharge directly to the aquifer by any suitable recharge techniques or by storing in storage then recharge daily. For direct recharge, the quantity of recharged amount would be $\frac{1964.31}{120} = 16.37\ \frac{m^3}{day}$, and *for daily recharge,* the quantity of recharged amount would be $\frac{1964.31}{365} = 5.38\ \frac{m^3}{day}$.

Source II: Harvestable rainwater from the building and open paved surface

Both buildings = 1223.83 + 740.48 = 1964.31 m^3.

Open paved surface = 707 × 2.9181 × 0.7 = 1444.17 m^3.

Total = (1964.31 + 1444.17) m^3 = 3408.5 m^3.

The total amount could be recharge directly to the aquifer by any suitable recharge techniques or by storing in storage then recharge daily. For direct recharge, the

RP: Recharge pits; RW: Recharge well; OW: Observation well; RT: Recharge trench; DW: Dug well; RS: Recharge shaft

Fig. 6.15 Managed aquifer recharge in the urbanized area

quantity of recharged amount would be $\frac{3408.5}{120} = 28.4 \frac{m^3}{day}$, and *for daily recharge*, the quantity of recharged amount would be $\frac{3408.5}{365} = 9.34 \frac{m^3}{day}$.

Source III: Only greywater [described in the Example problem 5.2].

For the seven-storied building, $32 \times 120 = 3840$ liters/day = 3.84 m^3/day

For the two-storied building, $9 \times 120 = 1080$ liters/day = 1.08 m^3/day

Total = (3.84 + 1.08) m^3/day = 4.92 m^3/day

The total amount would generate daily around the year, so direct recharge to the aquifer is recommended than storage.

The recharging option could be any single technique or combination of two or more methods are:

Option I: Recharge pit (RP)

Using typical pit of ($2m \times 2m$) at a depth of 3 m would have the capacity of 12 m^3. Thus, two recharge pits are suggested, as shown in Fig. 6.15.

Option II: Recharge well (RW) and observation well (OW)

The designed rechargeable water quantity is 33.32 m^3/day. There are two wells suggested, each of 100 mm diameter and 10 m long. The required gravel pack would design as per the available soil properties. One of the wells would act as an observation well (OW). Both wells are recommended to place at least 15 m away from the buildings (shown in Fig. 6.15).

Option III: Recharge trench (RT)

A single recharge trench is suggested instead of two recharge pits, as shown in Fig. 6.15.

Option IV: Dug well (DW) or dry well

If the area experiences dry shallow aquifer, there is a recommendation of two wells similar to option II.

Option V: Recharge shaft (RS)

As the total area is more than 1500 m^2, a shaft with a recharge well is suitable for groundwater augmentation.

Thus, detailed information on soil layers and the available groundwater table is essential to finalize the recharging technique.

References

ASCE (2020) Standard guidelines for managed aquifer recharge. ASCE/EWRI 69-19ISBN (print): 9780784415283ISBN (PDF): 9780784482179 (Standards). https://doi.org/10.1061/9780784415283

Auckland Council (2013) Stormwater disposal via soakage in the Auckland Region. TR2013/040.

Bettess R (1996) Infiltration drainage—manual of good practice. CIRIA Report 156.

CGWB (2007) Manual on artificial recharge of ground water. Government of India. Ministry of water resources. Central groundwater board.

Charlesworth PB et al (2002) The Burdekin Delta—Australia's oldest artificial recharge scheme. In: Dillon P (ed) Management of aquifer recharge for sustainability. Swets and Zeitlinger, Lisse, pp 347–352.

David R, Pyne G (2005) Aquifer storage recovery: a guide to groundwater recharge through wells, 2nd edn. ASR Press, Florida.

Gale I, Dillon P (2005) Strategies for managed aquifer recharge (MAR) in semi-arid areas. In: International hydrological programme (IHP), Paris.

Gammie G, Bievre BD (2015) Assessing green interventions for the water supply of Lima, Peru (Online Forest Trends). Accessed 09 May 2019.

Memola (2014) Historical/current water use (online). http://www.memolaproject.eu/sierra-nevada/wateruse. Accessed 08 Feb 2019

Page D et al (2011) Managed aquifer recharge case study risk assessments. CSIRO. https://doi.org/10.4225/08/5851899a5a3f

Stuyfzand P (2015) Managed aquifer recharge and storage (online). http://www.hydrology-amsterdam.nl/researchpages/Recharge.html. Accessed 24 April 2019

USEPA (2014) Underground injection control regulations. https://www.govinfo.gov/content/pkg/CFR-2014-title40-vol23/xml/CFR-2014-title40-vol23-part144.xml

USEPA (2016) National menu of best management practices (BMPs) for stormwater. National pollutant discharge elimination system.

Wang W et al (2014) Development of managed aquifer recharge in China. Bol Geol Min 125(2):227–233.

Weeks EP (2002) A historical overview of hydrologic studies of artificial recharge in the U.S. Geological survey. In: U.S. geological survey artificial recharge workshop proceedings, Sacramento, California, 2–4 April 2002, USGS Open-File Report 02-89.
Zhang H et al (2020) A review of the managed aquifer recharge: Historical development, current situation and perspectives. Phys Chem Earth Parts A/B/C 118–119:102887. https://doi.org/10.1016/j.pce.2020.102887

Chapter 7
Economics of Rainwater Harvesting System

7.1 Introduction

The economic analysis of a rainwater harvesting system depends on minimizing urban stormwater runoff and consumption of potable water and fuel/energy costs for treatment and distribution. The mode or design consideration of the rainwater harvesting system is demand-based. Economic assessment is required of any scheme regarding annual savings due to a rainwater harvesting system for water supply, greywater and urban runoff reduction, and managed aquifer recharge. Economic assessment for the rainwater harvesting system has been described in detail. The importance of the scientific evaluation of a rainwater harvesting system is to assess the benefit–cost ratio of the desired design. Before implementation, architectural plans could be evaluated besides the construction and maintenance costs involved for the engineering structures.

7.2 Life Cycle Cost Analysis

The Life Cycle Cost (LCC) analysis estimates the money needed on a scheme over its beneficial tenure. As per the Australian standard (1999), this is the sum of a rainwater harvesting project's procurement cost and ownership cost during its life cycle. LCC compares the costs and benefits for a rainwater harvesting investment, where the flows are discounted to net present equivalent values. Thus, LCC considers local conditions, i.e., annual rainfall, rainwater catchment configuration, harvested water demands, reduction of stormwater and greywater, local water prices with or without treatment, and potentialities for Managed Aquifer Recharge (MAR).

© The Author(s), under exclusive license to Springer Nature Switzerland AG 2022

A. Akter, *Rainwater Harvesting—Building a Water Smart City*, Springer Water,
https://doi.org/10.1007/978-3-030-94643-2_7

7.2.1 Costs

The costs of water collected in a rain tank depend on rainfall pattern, evapotranspiration (ET), LULC, catchment sizes, tank size, harvested rainwater demand, indoor use demands, and water price. Estimates of water and fuel savings are transferred into money (viz. dollar figures) based on the available cost spent. Here, energy savings on the harvested water usage are estimated through the pumping requirements. Capital costs include the storage tank and its fixtures, water supply network, and the pump. The price for a pipe network comprises construction method, diameter, material, depth, LULC, dewatering, and scale factor (i.e., length of pipes constructed under a single rainwater harvesting project). Usually, the installation cost represents 25% of the pump cost. While MAR uses harvested rainwater, surface runoff, and reusable greywater, the capital costs include:

- Land acquisition;
- Feasibility study or field tests;
- Consultation fees on design, local permission, and construction supervision;
- Construction costs include access road, conveyance, utilities, and treatments; and
- Routine test facilities for construction and operation.

Operational and maintenance costs are:

- Manpower for the system operation, regulatory, and administration;
- Energy consumption;
- Routine consultation allowance;
- Water quality testing facilities
- Maintenance for parts, storage tank, and fittings; and
- Treatment costs (e.g., reagents).

7.2.2 Benefits

The primary financial benefit minimizes the annual water bill from local water authorities. The yearly revenue is computed as the savings due to the harvested rainwater instead of the main water. The benefits are calculated from water supply and sewage system costs, as the supplied water ends up in the sewage system.

A cost–benefit study summarizes the overall relationship between the relative costs and benefits of a rainwater harvesting project.

7.2.3 Net Present Value (NPV)

The Net Present Value (NPV), a financial indicator, is the sum of *cash flows* discounted at a given rate. Cash flow refers to the money movements into and out of investment for a rainwater harvesting system in each period. Investors must implement the harvesting system and update the cash flows corresponding to a given discount rate during this period. The discount rate is a function of the interest rate (i), and the cash flow occurred in (t) years, as shown below:

$$Discount\ rate = \frac{1}{(1+i)^n}$$

The NVP has been obtained using the following expression:

$$NVP(i, N) = \sum_{t=0}^{N} \frac{(CF)_t}{(1+i)^t} \tag{7.1}$$

where
CF = Cash flow
i = Interest rate/discount rate
t = Project duration (years).

7.2.4 Internal Rate of Return (IRR)

The Internal Rate of Return (IRR) is the maximum return rate of a rainwater harvesting project, not more than the update rate after the project life. The IRR estimates potential investments' profitability and a discounted cash flow analysis having NPV of all cash flows equal to zero. The following expression can be applied to determine the IRR:

$$NVP = \sum_{t=0}^{N} \frac{(CF)_t}{(1+IRR)^t} = 0 \tag{7.2}$$

7.2.5 Payback Period (PB)

The Payback Period (PB) is expected to earn net revenue equal to a rainwater harvesting project's capital cost and calculated within the discount period. It is the ratio between total capital costs and the difference between the yearly revenue and expenditures (considering the discount rate). Thus, this approach overlooks the cash flows generated after recovering the investment and is not advised for long-term projects. Therefore, the payback period is:

$$PB = \frac{p + (CF)_p}{(CF)_p - (CF)_{p+1}} \quad (7.3)$$

p = Duration (years) immediately before the accumulated cash flow becomes positive

$(CF)_p$ = Cash flow in p duration

$(CF)_{p+1}$ = Cash flow in $p + 1$ duration

7.3 Economic Assessment of the System

Economic assessment defines as a process to identify, calculate and compare the benefit–cost of a project proposal either absolutely or in comparison with the alternatives. *The economic value of saved water* determines the value of potable water saved by implementing a rainwater harvesting system, i.e., direct market valuation or replacement cost. Similar to combined greywater reuse or MAR system. The rest of the rainwater has environmental significance while reduces urban floods or stagnant water. The *social cost* of reducing urban floods, groundwater depletion, and carbon emissions is based on the national or regional estimation. *Energy savings* are valued as per the available energy costs.

Economic assessments on rainwater harvesting are conducted through the benefit–cost analysis, cost-effectiveness analysis, and break-even analysis. *The benefit–cost analysis* quantifies in monetary terms of a specific rainwater harvesting policy proposal is feasible, including private, social costs and benefits, and items that are not entitled to receive a market-based adequate measure of economic value. *Cost-effectiveness analysis* is a partial benefit-cost approach that evaluates alternative same or similar outcomes. *Break-even analysis* determines the point at which the benefits of a policy option equal its costs. Through the break-even analysis, policymakers estimate the expected policy option to deliver the expected benefits. Among these three methods, cost-effectiveness analysis aims to achieve a specific outcome that is decided by the decision-makers. Benefit–cost analysis ensures the marginal benefit of the producing output equals the marginal cost. Break-even analysis is helpful to analyze policy options. Marginal cost and the marginal benefit are the first derivatives of the cost and the first derivative of the benefit to the output. For optimizing a multipurpose rainwater harvesting system, the above condition is presented as:

$$\frac{MB_1}{MC_1} = \frac{MB_2}{MC_2} = \ldots = \frac{MB_n}{MC_n} = 1$$

i.e.,

$$MB = MC$$

$$\frac{d(benefit)}{d(output)} = \frac{d(cost)}{d(output)} \tag{7.4}$$

The benefit–cost framework assesses the economic efficiency of harvesting in a particular area. Thus, the benefit–cost analysis compares the benefits of reduced potable water use with the costs of implementing and maintaining a rainwater harvesting system. Assumptions in the benefit–cost analysis are (Dallman et al. 2016):

A. No additional costs are required for the overall water use to adopt property owners' rainwater harvesting but offer a permanent decline in water utility costs.
B. Greywater reuse or stormwater runoff reduction benefits of rainwater harvesting system, i.e., water quality improvements or urban flood risk and damages are minimized due to reduced runoff peak and volume, considered in the benefit–cost analysis.
C. The benefits of harvested rainwater included are the economic value of water. Also, energy, fuel costs, and carbon emissions are saved. Harvested rainwater offers intrinsic satisfaction or personal pride for the property owner due to the water savings even if the higher prices exist than the monetary benefits.
D. The cost per cistern should be included.
E. Benefit per cistern in terms of rainwater harvesting system users.
F. Capital costs are required to purchase and install a rainwater harvesting system. Also, acquired benefits from the increased savings of water and energy consumptions and carbon emissions each year during the entire service life of the project.

The Net Present Value (NPV) computes the overall value of an option in benefit–cost analysis. The NPV is the difference between the present value of cash inflows and outflows over a specific duration. If the *NPV is positive*, the investment improves efficiency because it involves benefits over time, more than outweighs costs, i.e., assumptions D and E. If the NPV is negative, the proposal is inefficient (the costs outweigh the benefits), i.e., assumptions B and C. The size of their NPV can compare policy options. Using NPV, financial performance evaluation of rainwater harvesting system has been carried out considering housing type, tank size, and water price.

The economic assessment of a rainwater harvesting system has been conducted through the Net Present Value (NPV), the Payback Period (PB), and the Internal Rate of Return (IRR). Mathematically, these lead to either a maximum or minimum condition. Hence, to ensure the optimization being the maximum, the secondary condition supposes to be valid. Then, Eq. (7.4) should be:

$$\frac{d^2(benefit)}{d(output)^2} < \frac{d^2(cost)}{d(output)^2} \tag{7.5}$$

Theoretical analysis needs to be conducted for economic division, independent rather than hydrologic and financial uncertainties. Therefore, a modified decision has been required based on expertise and judgment on intangible factors.

The *Minimum Acceptable Rate of Return* (MARR) is the lowest rate of return on a project accepted by the owner before starting a project, given its risk and the opportunity cost of forgoing other projects, could be estimated as follows:

$$MARR = Inflation + risk \tag{7.6}$$

$$NPV = \frac{Cash\,flow}{(1+i)^t} - initial\ investment \tag{7.7}$$

t = number of periods
i = required return or discount rate; set as MARR

7.4 Integrated Rainwater Harvesting and Others

Integrated rainwater harvesting systems with either greywater reuse or aquifer recharge technologies are assessed through:

Step 1: Investment and operation costs: The investment is needed for rainwater harvesting systems, and greywater reuse should be estimated with or without treatment systems. Additionally, the operation and maintenance costs are required to be assessed.

Step 2: Benefits: The economic benefits are obtained by implementing the integrated system in potable water savings and wastewater reduction. The benefits extend to the potable water savings, the decrease in pumping costs, groundwater or treated surface water savings, and minimize wastewater treatment.

Step 3: Cash flows and metrics: Considering investments, operation and maintenance costs, and estimated benefits, the cash flows are prepared.

Example problem 7.1 An example has been adapted from Matos et al. (2015) for a commercial building, Dolce Vita Braga shopping center in the Braga, which existed north of Portugal. There are several distinct but complementary areas, i.e., shopping areas with commercial and retail area spaces, restaurants, leisure areas, and supermarkets. The intervention area available for the project is 159,971 m^2, and the footprint of the whole commercial area is 46,611 m^2, for a gross floor area of 90,000 m^2. Structurally, the building is distributed on three floors. The retail units are on a single floor, supported by public parking spread over four floors. Overall, the business has a gross leasable area of 75,000 m^2, corresponding to 165 units allocated to different activities. In addition, there is a total parking area of 62,000 m^2, distributed over four basements and outer surface parking places, corresponding to 2,750 m^2 car parking spaces. Considering non-potable uses for rainwater harvesting, sizing the rainwater storage tank of commercial area with an extensive collection roof surface, considering the different possibilities of rainwater use and additional time interval consideration, i.e., one year and seven months.

Table 7.1 Financial performance of the rainwater harvesting systems in residential buildings

Country	Financial performance	
	Positive NPV	Negative NPV
UK (Ward et al. 2010)	Good design	Poor design
Spain (Guisasola et al. 2011)	• Group of houses • Group of apartment buildings	• Two single houses • Eight single houses • Apartment houses • Apartment buildings
Granollers, Catalonia, Spain (Guisasola et al. 2011)	• High water price	• Low water price
Greater Sydney, Australia (Hajani et al. 2013)	Tank size 5000 liters	Tank size 2000 and 3000 liters

Table 7.2 Summary of the estimated savings (Matos et al. 2015)

End-uses	Washing parking floors and garden irrigation
The volume of the storage tank (@1-year interval) (m^3)	7,277.29
The volume of the storage tank (@7 months interval) (m^3)	11.63
Consumption (m^3/month)	3,302. 20
Saving in public drinking water (USD/month)	5849.13

The storage tank and its fixtures, i.e., the alternative water supply and the pump, are considered investments. The project owner required that the rainwater harvesting tank be undergrounded to improve its installation's environmental visual impact. The cost of the horizontal polyethylene rainwater harvesting tank and its fixtures were based on their total volume. These companies have indicated that the cost of the fixtures is typically 30% of the reservoir.

The discount rate (i) used for benefits-costs studies in Portugal is 6%, not including the inflation rate as per European Commission, Directorate General Regional Policy (European Commission 2008); for this work, the discount rates of 5% and 10% were used to take in account some uncertainties. And the evaluation period (t) is 20 years.

Considering Tables 7.1, 7.2, 7.3, 7.4 and 7.5 *as the given data, conduct an economic assessment of the rainwater harvesting system.*

Solution:

Table 7.6 presents the economic assessment of the rainwater harvesting system, with an interest rate of 10%, for a discount rate of 10%, the expected NPV found as 282 MUS$. The PB is approximately one year and a half. Suppose a discount rate of 5% is considered the expected NPV is 437 MUS$. The PB is two years; the water savings have been achieved by 18–20%. This benefit–cost analysis considers main water supply savings as the only benefit of a rainwater harvesting system. Here the maintenance costs are absent. Also, the environmental, social, financial, and energy-saving benefits have been ignored.

Table 7.3 Water tariff for non-domestic consumption in the municipality of Braga (Matos et al. 2015)

Classes (m^3)	USD/m^3
1° Class—0 to 30	1.065
2° Class—31 to 60	1.513
3° Class—>60	1.670
Rate of water resources (USD/m^3)	
Sanitation	0.012
Water	0.025
Rate connection sanitation (USD/month)	
Building not intended for housing (an area greater than 100 m^2)	4.707

Table 7.4 Costs of the elements included in the rainwater harvesting infrastructures

Item	Cost expressed in USD
Storage	
Pre-fabricated tank	
Polyster tank with filter (including excavation)	6316.2
Estimation accessories (30% of the cost of the tank)	1894.86
Distribution	
Pumping station pump	10,043
Estimation of the cost of installation (25% of the cost of installation)	2510.75
Distribution system	
Cleaning of floors of parking lots	
D32 mm [m] @ 300	2395.8
D125 mm [m] @ 520	30,012.84
Garden irrigation	
D32 mm [m] @ 1800	14,374.8
D65/50 mm [m] 1000	23,135.2
Total	90,683.45
Taxes (23%)	111,465.2

7.5 Multi-Criteria Analysis (MCA)

The Multi-Criteria Analysis (MCA) technique assesses policy options considering quantitative and qualitative impacts. The approach includes broader criteria (viz., social and environmental factors), need to measure in the most relevant unit instead of monetary values. MCA fails to inform the decision-maker on an individual proposal to achieve the net social benefit. The details on advancement in MCA has described in Chap. 8.

Table 7.5 Application of the water tariff for non-domestic consumption in the municipality

Components	Quantity (m^3)	Price (USD/m^3)	Total (USD)	Taxes (6%)	Total + taxes (USD)
Pavement washing classes					
1°Class—0 to 30	30.00	1.06	31.94	1.91	33.86
2°Class—31 to 60	30.00	1.51	45.38	2.72	48.10
3°Class—> 60	1,240.00	1.67	1970.36	118.22	2088.58
Rate of water resources (USD/m^3)					
Sanitation	1,240.00	0.01	15.15	0.91	16.07
Water	1,240.00	0.02	30.75	1.85	32.61
Rate connection sanitation (USD/month)					
Building not intended for housing (area greater than 100 m^2)	1.00	4.71	4.71	0.28	4.99
				1 month	2098.30
				7 month	14,688.12
				1 month	2224.20
				7 month	15,569.40
Garden irrigation classes (USD/m^3)					
1°Class—0 to 30	30.00	1.06	31.94	1.91	33.86
2°Class—31 to 60	30.00	1.51	45.38	2.72	48.10
3°Class—>60	2,002.20	1.67	3343.28	200.59	3543.87
Rate of water resources (USD/m^3)					
Sanitation	2,062.20	0.01	25.20	1.51	26.72
Water	2,062.20	0.02	51.16	3.07	54.22
Rate connection sanitation (USD/m^3)					
Building not intended for housing (area greater than 100 m^2)	1.00	4.71	4.71	0.28	4.99
				1 month	3501.66
				7 month	24,511.59
			+ taxes	1 month	3711.76
				7 month	25,982.28
		Total		1 month	5598.74
				7 month	39,199.69
		Total + Taxes		1 month	5935.96
				7 month	41,551.68

Table 7.6 Economic assessment of the rainwater harvesting with an interest rate of 10%

	Interest rate i = 10%		Present cash flow (USD)	Payback period (PB) = 1 year 5 months and 26 days		Accumulated cash flow (USD)
Year	Investments/ costs	Benefits		$\frac{1}{(1+i)^n}$	Updated cash flow	
0	90,683.45	39,199.69	51,483.76	1.0000	51,483.76	51,483.76
1		39,199.69	39,199.69	0.9091	35,636.09	15,847.67
2		39,199.69	39,199.69	0.0826	32,396.44	16,548.77
3		39,199.69	39,199.69	0.0751	29,451.32	46,000.08
4		39,199.69	39,199.69	0.6830	26,773.92	72,773.99
5		39,199.69	39,199.69	0.6209	24,339.92	97,113.92
6		39,199.69	39,199.69	0.5645	22,127.21	119,241.12
7		39,199.69	39,199.69	0.5132	20,115.65	139,356.76
8		39,199.69	39,199.69	0.4665	18,286.95	157,643.71
9		39,199.69	39,199.69	0.4241	16,624.49	174,268.21
10		39,199.69	39,199.69	0.3855	15,113.18	189,381.38
11		39,199.69	39,199.69	0.3505	13,739.26	203,120.64
12		39,199.69	39,199.69	0.3186	12,490.24	215,610.87
13		39,199.69	39,199.69	0.2897	11,354.76	226,965.62
14		39,199.69	39,199.69	0.2633	10,322.51	237,288.13
15		39,199.69	39,199.69	0.2394	9,384.09	246,672.22
16		39,199.69	39,199.69	0.2176	8,531.00	255,203.22
17		39,199.69	39,199.69	0.1978	7,754.89	262,958.67
18		39,199.69	39,199.69	0.1799	7,050.42	270,009.08
19		39,199.69	39,199.69	0.1635	6,409.47	276,418.54
20		39,199.69	39,199.69	0.1486	5,826.78	282,245.32

References

Australia Standards (1999) Australian/New Zealand Standard™. As/nzs 4536:1999 life cycle costing-an application guide, reconfirmed 2014. https://www.saiglobal.com/online/. Accessed 23 Jan 2021.

Dallman S et al (2016) The value of rain: benefit-cost analysis of rainwater harvesting systems. Water Resour Manage 30(12):4415–4428. https://doi.org/10.1007/s11269-016-1429-0

Ward S et al (2010) Harvested rainwater quality: the importance of appropriate design. Water Sci Technol 61(7):1707–1714. https://doi.org/10.2166/wst.2010.102

Guisasola A et al (2011) Roof selection for rainwater harvesting : quantity and quality assessments in Spain 5. https://doi.org/10.1016/j.watres.2011.03.036

Hajani E et al (2013) Reliability analysis for rainwater harvesting system in peri-urban regions of Greater Sydney, Australia. In: Adapting to change: the multiple roles of modelling: proceedings of the 20th internationa congress on modelling and simulation.

Matos C et al (2015) Economic analysis of a rainwater harvesting system in a commercial building. Water Resour Manage 29(11):3971–3986. https://doi.org/10.1007/s11269-015-1040-9

European Commission (2008) Guide to cost-benefit analysis of investment projects. https://ec.europa.eu/regional_policy/sources/docgener/guides/cost/guide2008_en.pdf

Chapter 8
Advanced Technologies in the Water Smart City

8.1 Introduction

Rainwater harvesting for the water smart city faces dynamic challenges ranging from land use land cover (LULC) and hydrological cycle alteration. Thus, the planner and designer need advancement in the long term as well as real-time dataset handling. Then, these datasets engaged either in the mathematical or numerical model to feature the rainwater harvesting system for the design purpose, i.e., potable water supply, stormwater management, integrated with greywater management, or/and Managed Aquifer Recharge (MAR). Once the water resource modeler declares the water demand and rainwater supply, construction works can be progressed in the next phase, followed by the recommendations from socio-economic modelers. This chapter described the advancements in the hydrologic study, demand-based models on rainwater harvesting systems, automated rainwater treatment, advanced tools for planners and architects, and socio-economic models. Compared to the conventional practices for hydrological study, data integrity could be achieved using advanced technologies. Rainwater harvesting system component-wise operation models are defined in Chap. 3. However, dedicated techniques/models to meet the users' demand have been included in this chapter. In addition, an automated rainwater treatment case study has been incorporated. Tools for the planners and architects to determine the potentiality and feasibility of the rainwater harvesting system are described. Lessons from the existing rainwater harvesting system provide insight, socio-economic models have been discussed to emphasize the decision support criteria for the new rainwater harvesting system. Thus, the decision support system would benefit by adopting advanced active techniques by establishing a dynamic interaction between rainwater harvesting and the water smart city.

© The Author(s), under exclusive license to Springer Nature Switzerland AG 2022

A. Akter, *Rainwater Harvesting—Building a Water Smart City*, Springer Water,
https://doi.org/10.1007/978-3-030-94643-2_8

8.2 Advancement in the Study of Hydrology

Performances of rainwater harvesting are primarily dependent on rainfall variability, planning, and designing. Thus, the recent advanced technologies in real-time rainfall data acquisition and prediction play a vital role in designing.

8.2.1 Remote Sensing

The remote sensing technique can observe global precipitation compared to the ground rain gauges stated in Chap. 2. The working principle of remote sensing is based on extrapolating surface parameters after the device recorded upwelling electromagnetic radiation from the ground surface. There are two primary precipitation observation techniques, i.e., Weather radars and Space-based meteorological satellites. After World War II, *weather radars* obtained huge developments with high Spatio-temporal resolutions to record precipitation (Fabry et al. 1994; Morin et al. 1995; Harris et al. 2001; Berne et al. 2004). The electromagnetic energy sources are usually installed on satellites, and these sensors acquire data from the earth. The image processing technique is applied to process digital images.

Developing algorithms and models for assessing hydrometeorological data is the continuous latest remote sensing technique compared to the existing information on the field, vegetation, and watershed up to regional scales. The variables are the land surface temperature, near-surface soil moisture, water quality, landscape roughness, and LULC. Thus, applying remote sensing techniques in hydrology could supplement hydrometeorological states and fluxes estimation (Schmugge et al. 2002). In this regard, the hydrometeorological fluxes include evapotranspiration and snowmelt runoff.

8.2.2 Geographical Information System

The Geographic Information System (GIS) is a computer software system comprising hardware and data, merged with expertized handling to support manipulating, analyzing, and displaying information in a spatial location. A GIS is usually a "smart map" containing related features derived from a database. Thus, GIS combines geographic coordinates with attributes for characterizing any geo-referenced data describing natural or artificial/urbanized phenomena. GIS can offer a wide range of functions to access data, geographic databases management, analyze/model, and display output.

Applications of GIS in hydrology are advantageous for catchment-based assessment, analysis of surface and groundwater, both quality and quantity context. GIS integrates multiple sources, including satellite imagery, topography, LULC, boreholes and wells, subsurface isopach maps, and surface geology. Therefore, an understanding could be developed on surface and subsurface water movement and their interactions. Continuous information offers critical insight into the detailed physical hydrological processes available compared to the intermittent manual measures. Many of the satellite-based data are either free or required negligible costs from government entities. For example, using GIS platform USGS Digital Elevation Models (DEMs), USGS Digital Orthophoto Quarter Quadrangles (DOQQs), and Landsat images present visual information on the surface or subsurface processes. Then, DEM has derived from either contour lines or photogrammetric methods. DEM offers the digital cartographic dataset in three coordinates (i.e., XYZ). In this regard, the terrain elevations are recorded from the ground positions at regular horizontal intervals.

8.2.3 Artificial Neural Networks

Artificial Neural Networks (ANNs) replicate human brain functions through a learning process and obtain knowledge on optimal weights for the connections and threshold values for the nodes. The basic steps involved within the network are:

(a) Information sharing between nodes through connection links;
(b) Signals sharing between nodes through connection links;
(c) Each connection link bear an associated weight to represent the connection and strength; and
(d) Each node applies a non-linear transformation, i.e., activation function, to determine the output signal of the network.

ANNs are robust modeling tools for handling nonlinear hydrologic processes, including rainfall-runoff, streamflow, groundwater flow, and water quality simulation. Followed by appropriate training, a reasonable prediction could be achieved for many hydrological processes. The input vector could enrich a detailed understanding of the hydrologic process and design a relatively efficient network. Apart from long-term forecasting, the ANNs apply to forecast short-term rainfall to operate in real-time.

8.2.4 Genetic Algorithm

A Genetic Algorithm is based on natural genetics and natural selection to develop search and optimization procedures. The working principles are the survival of the fittest and the inheritance of the characteristics of the parent populations. A genetic algorithm quickly solves the problems associated with the non-convex functions compared to the conventional optimizations. A genetic algorithm requires converting design space into genetic space, but this is simultaneously a process of several solutions. The following steps are involved in the genetic algorithm:

(a) Generates an initial population;
(b) Refer to a coding scheme for all the variables;
(c) Flow simulation finite element modeling for the variables;
(d) Processing fitness from objective functions;
(e) Performing using genetic operators; and
(f) Termination condition.

8.2.5 Fuzzy Logic

Fuzzy logic can solve imprecise or vague problems in artificial intelligence. When the information is incomplete and ambiguous, precise mathematics seems insufficient to model a complex system, and fuzzy logic can solve these. Mathematically, a fuzzy could be described by assigning each possible individual its grades of membership in the fuzzy set. These individuals fit the fuzzy set either in a greater or lesser degree, i.e., membership grade, and often expressed by actual numbers between 0 and 1. A fuzzy system acts in modeling, data analysis, prediction, or control. *Fuzzy* theories effectively handle dynamic, non-linear, and reasoning noisy rainfall data. The fuzzy logic has excellent potential for weather forecasts, including long-term rainfall forecasting.

8.2.6 Rainfall Mapping

Rainfall mapping comprises both the spatial and temporal variability of rainfall. Inland surface hydrology, spatial rainfall dataset includes areal rain over a region and transforms these data into a *rainfall map*. The temporal rainfall dataset is the long records of daily or hourly precipitation recorded from rain-gauge (stated in Chap. 2). Preparing rainfall maps from temporal data as well as satellite-based images are followed by spatial interpolations using GIS. A typical rainfall map generates by combining temporal and spatial datasets (Fig. 8.1). Here, the rainfall anomaly analysis compares rainfall data between multi-satellite precipitation analysis data and gauged-based products.

Based on the interpolation method and application scale, there are three categories (Hutchinson 1998; Friedman et al. 1991; Tomczak 1998; Cheng et al. 2009; Huang et al. 2010; Alan and Ali 2011; Wong et al. 2016).

- Category I (interpolating temporal rainfall data): this is a relatively simple method and is preferable for small to medium-scale catchments or basins. These interpolations include Nearest Neighbour (NN), Thiessen polygons, Spline, and various Kriging and Inverse Distance Weighting (IDW). *Nearest Neighbour (NN)* is an

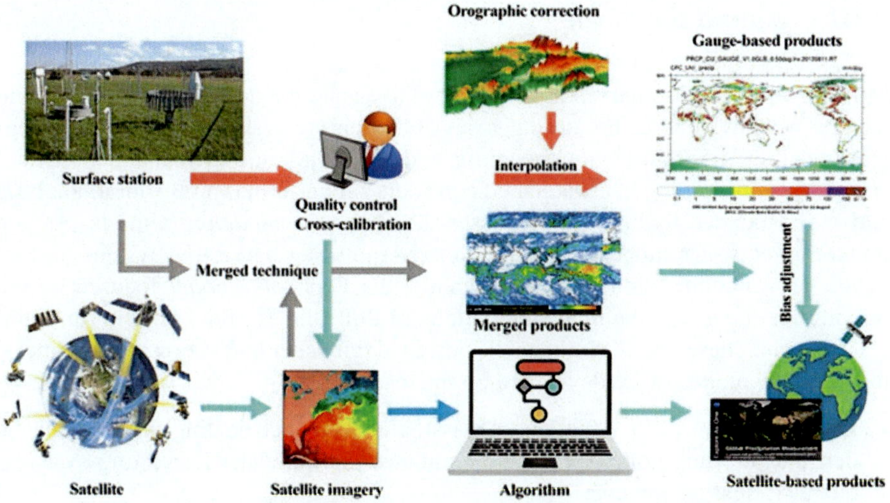

Fig. 8.1 Satellite-based precipitation products. The images for satellite adapted from Sun et al. (2018)

analog-type approach based on the statistical downscaling to derive local-scale information of precipitation from numerical weather prediction model output. *Inverse Distance Weighting (IDW)* is a deterministic multivariate interpolation method that uses an available scattered point rainfall dataset. The assigned values to the unknown rain are estimated with a weighted average of available rain values. A thin plate smoothing *spline* works on the dataset allowing errors in each data point, and smoother outcomes are obtained from the finer dataset. Thus, spline interpolations on daily rainfall are based on a mean annual rainfall surface instead of elevation. *Kriging* disintegrates the stochastic process that accumulates daily rain from a linear trend and a stochastic error process.

- Category II: This category applies to the interpolation process for predicting large-scale rainfall using ancillary data, viz. satellite imagery and DEMs, along with temporal data from rain gauge stations.
- Category III: to forecast rainfall, this category applies complex interpolation using fuzzy reasoning and ANNs.

8.3 Application of Rainwater Harvesting Modeling Software

Application of software to evaluate the performance of rainwater harvesting systems considers the hydrological characteristics, environmental, social, and economic consequences.

8.3.1 Around the World

Modeling of rainwater harvesting is required to ensure the desired water supply and quality. Worldwide three broad categories of models are available, i.e., (i) hydrological modeling, (ii) optimization modeling, and (iii) economic and financial modeling. *Hydrological modeling* in a watershed considers a long-term rainfall-runoff analysis and the associated hydrologic processes. *Optimization modeling* aims to provide uninterrupted water supply through optimized rainwater harvesting system components, i.e., catchment and rain tank. Finally, the *Economic model* focused on the initial investment, and the financial modeling fulfills the essential criteria for financial analysis. Therefore, to simulate the designed rainwater harvesting performances, the available models were developed by the researchers:

- DRHM (Dixon et al. 1999) is a mass-balance model dealing with stochastic demand profiling components and simulates the rainwater harvesting system's quantity, quality, and cost.
- Rewaput model (Vaes and Berlamont 2001), a reservoir model based on the IDF relationship and a triangular distribution, assess the stochastic character of the rain tank's capacity and water consumption (i.e., vary within a catchment).
- RSR model (Kim and Han 2006) minimizes stormwater and flooding by optimizing a rainwater harvesting system's tank size.
- RCSM (Fewkes 2000) models rainwater harvesting systems using time-interval variations and yield before or after the spill.
- RainCycle (Roebuck and Ashley 2007), an excel-based balance model, deals with a yield after the spill algorithm and an entire life cycle cost. RainCycle complies with current UK best practice guidance. This model is based on the YAS algorithm (described in Chap. 3).
- SimTanka, modeling rainwater harvesting systems, was formulated in 2019 by the Vikram Vyas the Ajit Foundation, India Habitat Centre, New Delhi.
- OptiRTC, the rainwater harvesting controller, works based on the software and online weather forecasts. Here, internet-based weather forecast data automatically empties rainwater systems before storm events to maximize storage and reduce impacts on the stormwater system.
- For analyzing rainwater harvesting systems, there are some models to aid design and performance. RWIN (Herrmann and Schmida 2000) is based on the hydrological-based precipitation runoff model simulation for rainwater harvesting system evaluation.
- The rainwater TANK model (Vieritz et al. 2015) was developed for urban households in Queensland, Australia. Presently, rainwater TANK is a continuous simulation daily time-step model that calculates the water balance of the roof, rainwater tank, internal water use areas, and one external water use. The model incorporates a first flush device. If the model indicates that spillage is excessive, the user can increase the capacity of the tank store and rerun the model, and iteratively arrive at a suitable design.

Building Information Modeling (BIM) is a leading tool to illustrate and pre-execute an overview of a city's physical and functional characteristics. Thus, features of BIM tools become helpful for future planning using drone-based surveying, 3D printing, and advanced transportation system towards water smart building management.

8.3.2 Principles of Modeling Software

As the water smart city broadly meets the requirements of the smart city (stated in Chap. 1, art 1.3), the modeling software should also maintain the fundamental four pillars, including social, physical, institutional, and economic. The modeling software principles are:

Principle 1: Model selection should be traceable to the requirements of the rainwater harvesting system.
Principle 2: Model setup should follow the planning and architecture of the proposed rainwater harvesting system.
Principle 3: Hydrological, optimization, economic and financial models are of equal importance; and
Principle 4: Model validations play a vital role.

8.3.3 Possibilities for Stormwater Management

A water smart city combines planning and management in the urban design, protection, and urban water cycle conservation to mimic natural hydrological and ecological processes. Conventional urban development significantly impacts the natural process by altering the hydrologic cycle and transmitting stormwater runoff to waterways. As the conveyance, treatment, storage, and distribution of stormwater are expensive and time-consuming. Therefore, the application of several numerical models (Table 8.1) has been initiated worldwide to evaluate the conceptual design of stormwater management to meet the desired water quality. A comparison has been drawn in terms of the on-site detention or retention, rainwater storage, pervious pavements, buffer strips, bioretention swales, sedimentation basins, ponds, constructed wetlands, infiltration systems, gross pollutant traps, oil and grit separators, stormwater harvesting, rain garden, green roofs, street sweeping, onsite wastewater management, community wastewater management schemes, and demand reduction.

Table 8.1 Comparative performances of stormwater models (Rossman 2006)

	MUSIC	EPA SWMM	XP-SWMM	Watercress	Drains	Hec-RAS	SWITCH	Switch2	PermPave	Raintank analyser	E2
On-site detention	√	√	√	√	√	×	×	√	√	×	×
Onsite retention	√	√	√	√	×	×	√	√	√	×	×
Rainwater storage	√	√	√	√	×	×	×	√	×	√	√
Pervious pavements	√	×	×	×	×	×	√	√	√	×	×
Buffer strips	√	×	×	×	×	√	×	×	×	×	√
Swales	√	√	√	×	×	√	×	×	×	×	√
Bioretention systems	√	√	√	×	√	√	×	√	×	×	√
Sedimentation basins	√	√	√	×	√	√	×	×	×	×	√
Ponds	√	√	√	×	×	√	×	×	×	×	√
Constructed wetlands	√	√	√	√	√	√	×	×	×	×	√
Infiltration systems	√	√	√	√	√	√	×	×	×	×	√
Gross pollutant traps	√	√	√	√	×	√	√	√	√	×	×
Oil and grit separators	×	×	×	×	×	×	×	×	×	×	×

(continued)

Table 8.1 (continued)

	MUSIC	EPA SWMM	XP-SWMM	Watercress	Drains	Hec-RAS	SWITCH	Switch2	PermPave	Raintank analyser	E2
Stormwater harvesting	√	√	√	√	×	×	×	√	√	√	√
Rain Garden	√	×	√	√	×	×	×	√	×	×	×
Green roofs	×	×	√	×	×	×	×	×	×	×	×
Street sweeping	×	×	√	×	×	×	×	×	×	×	×
Onsite wastewater management	×	×	×	×	×	×	×	×	×	×	×
Community wastewater management schemes	×	×	×	×	×	×	×	×	×	×	×
Demand reduction		×	×	√	×	×	×	√	×	×	√

√ computes by the concern model
× the concern model unable to compute

8.3.4 Possibilities for MAR

In the Managed Aquifer Recharge (MAR), rainwater or recycled water has considered to be routed into the subsurface. Characteristics or parameters of the geologic medium are known. The 'physically consistent' model describes the water movement and pollutants intrusion through a geological porous medium. Thus, the mathematical modeling of regional aquifers applies:

- to predict the changed response of the aquifer due to urbanization, i.e., new pumping, infiltration shifts, over-extraction for irrigation, pollutant contaminations;
- to offer the required information compatible with the local regulations;
- to obtain a better understanding of the aquifer system in the geological, hydrogeological, and hydrochemical context; and
- to disseminate knowledge on the improved observation networks and field experiments.

The aquifer characteristics, i.e., hydraulic conductivity, effective porosity, disparities', etc., are 'equivalent' or averaged values on the Representative Volume Element (RVE). The hydrogeological parameters are acquired through a calibration process applied on the entire aquifer, verified through the best agreement between the observed and the calculated values. Various values are tested for the selected parameters considering local heterogeneities. Therefore, the accuracy of the simulation is solely influenced by the scale of the issue.

The long-term performance of combined MAR or SAT solutions for the operational implementation could be assessed by advanced monitoring and modeling. *AquaNES* is based on integrating nature-based elements solutions towards urbanization challenges. These are inspired and supported by natural processes, thus considering environmental and socio-economic objectives becomes a sustainable measure. For assessing protection zones around the main pumping wells, a detailed groundwater transport modeling involves:

- the local calibration on flow-based model considers measured piezometric levels;
- the local transport calibration depends on the observed breakthrough curves from tracer tests or measured contaminations);
- the simulations of travel times for various injection points.

The numerical modeling of a regional aquifer carried out through:

- Existing data collection;
- Data preparation and model setup;
- New measurements campaigns;
- Local-scale model simulation; and
- Local-scale groundwater quality study.

The numerical models ensure optimization of the management and groundwater resources protection. These simulations offer sustainable groundwater resources, represent a physical and environmental decision support system for the decision-makers.

8.4 Suitable Site Selection for Rainwater Harvesting

The best site selection for harvesting rainwater usually considers the socio-economic factors, i.e., set back to road and settlements, and physical characteristics (i.e., terrain slope, soil types, and LULC) of the city/urban area.

8.4.1 Application of Drones

Recent rapid progress in the affordable production of drones and Unmanned Aerial Vehicle (UAV) technology has shown that the drone platform can effectively use survey and monitoring applications. The drones can have installed Red–Green–Blue (RGB) or hyperspectral cameras and Light Detection and Ranging (LiDAR) equipment to acquire high-resolution data from the land surface. The drone platform can also include pre/post-processing software to handle the received data, enabling users to produce 3D point data for real-time simulations. These recent developments, i.e., advancements in remote sensing, drone platform development, measurement by LiDAR, and hyperspectral imaging processing, provide details on rainfall consequences. Field experiments and data analysis within the data-intensive models must be performed to validate urban stormwater runoff and groundwater recharge.

8.4.2 GIS and Remote Sensing

Geospatial technologies, i.e., GIS associated with remote sensing, are practiced for site identification designed for a rainwater harvesting system and MAR zone. Similarly, GIS has been a practical tool for assessing MAR zones and suitable rainwater harvesting system sites for constructing structures based on modern scientific principles to ensure sustainability. Available GIS-based remote sensing studies combine various factors, including lineament density, drainage density, slope, soil permeability, LULC, geology, geomorphology, urban flooding, groundwater level, etc. (Krishnamurthy et al. 2000; Sener et al. 2005; Shaban et al. 2006; Solomon and Quiel 2006; Tweed et al. 2007; Riad et al. 2011; Pokhrel et al. 2012). Additionally, determining a suitable site or zone for harvesting rainwater is based on a weighted overlay process.

8.5 Rainwater Harvesting with Architectural Design

Rainwater harvesting has become an integral part of the architectural design to build a water smart city. For example, Le Corbusier's urban masterplan Ville Radieuse (The Radiant City) in 1933 has aimed at the future or proposed planning for prefabricated and identical high-density skyscrapers distributed across a considerable green area. These components are organized in a Cartesian grid towards functioning the city as a "living machine". In addition, to retrieve the sunlight green space lost beneath the skyscrapers, Le Corbusier proposed extensive roof gardens. Also, Le Corbusier suggested constructing elevated highways and auto-ports 5 m above the ground level to reserve the entire ground level for pedestrians (Scarlett 2021).

Built form in architecture refers to the function, shape, and configuration of buildings as well as their relationship to adjacent streets and open spaces. Thus, the geometric variables of a building are: shape factor (i.e., the ratio of building length to building height in the plan), building height, catchment type (stated in Chap. 3), catchment slope, facade slope, and the local climate responsive design to acquire optimal benefits as a part of the water smart city. Rainwater collection has become a part of architectural design and adds distinction while adapting to the individual or small buildings. Latest commercial infrastructures introduce stormwater management within the architectural plans to achieve the government financial benefits due to conservation. The architectural design suggests innovative roof pitches, gutters, and water tanks to store rainwater correctly.

Hydro-meteorological parameters consideration

- Solar radiation has been prioritized in temperate-humid climate regions to meet the more extended heating duration. In these regions, primary living areas of houses design to face east, south, and west. The building forms in these regions are more flexible and various.
- Dense settlements and low-rise buildings are designed to protect from sunlight and wind. However, humid winds are allowed for a significant duration in a year in the hot-dry climatic region.
- Low precipitation and flat roofs should be subjected to shade the adjacent streets throughout the day. In the hot region, the essential aspect includes proper paint selection for building and providing high boundary walls.

Roof patterns

- A 'V' shaped inverted roof proves efficient rainwater collection followed by filtration to a cistern for storage, for example: Herreros Arquitectos, Arta, Spain.
- The provision of a wooden home pitcher acts as a wind sail to ensure airflow in a single direction and convey rains for harvesting example: Fernanda Vuilleumier Studio, Puerto Natales, Chile.
- An aluminium overhang collects rainwater to comply with local plans and acts as a Japanese-style fountain to filter water towards a retention pool, for example, Avignon-Clouet Architecture, Rezé, France.

- A butterfly roof conveys rainwater to a home side cistern for potable use treatment conducted through an internal charcoal filter and ultraviolet light example: Sanders Pace Architecture, Sharps Chapel, US.
- The roof acts as a natural basin for rainwater harvesting. Thus, the harvested water utilizes electricity to heat pumps and radiant loops for the heating and cooling of the residence example: Cascading Creek House by Bercy Chen Studio LP, US.
- The hilltop houses are naturally ventilated and could practice sustainability by adapting energy-efficient LED lights and rainwater harvesting. Example: Richard Cole Architecture, Australia.
- This eco-friendly home's rainwater recovery is used for gardening and irrigation to aromatic plants without compromising the public water supply. Example: Eco-Sustainable Antony House by Djuric Tardio Architectes, Antony, France.

Thus, building designers and owners can showcase water conservation and incorporate aesthetics to promote a water smart city. Therefore, this practices customer and general interest in conservation and rainwater collection and develops public awareness of sustainable living practices. Thus, the following factors should be considered by an architect while designing a rainwater harvesting system:

- Floor Area Ratio (FAR) is the ratio between the total floor area and the plot area.
- Maximum Ground Coverage (MGC) is the maximum covered area of a plot by the building. The MGC of a building depends on plot size, land use pattern, and road width.
- The site set back area and the area beyond the allowable MGC is considered Mandatory Open Space (MOS).
- Site setback area must be kept open in front, on both sides, and in the back of the building.
- The area within the plot must be left as bare ground and open to the sky called Mandatory Unpaved Area (MUPA).

8.5.1 Computer-Aided Design (CAD) Software

Computer-Aided Design (CAD) applies in creating, modifying, analyzing, or optimizing a design. For example, AutoCAD® is a software that creates precise 2D and 3D drawings for architects, engineers, and construction professionals. Autodesk Revit® software facilitates a workflow-type Building Information Modelling (BIM) approach. BIM includes architectural and structural elements in designing rainwater harvesting systems.

On the other hand, Rhinoceros is one of the fastest-growing 3D modeling tools for architects and urban designers. Coupled with Grasshopper, a graphical algorithm editor, Rainwater + has been developed by the Harvard research group stated in Chen et al. (2016). This tool is applied for urban rainwater runoff assessment and management to serve architects, landscape architects, and urban designers.

8.6 Automated Water Quality

The properties of a catchment, storage, and distribution network significantly influence the quality of harvested rainwater. The first flush ensures quality during the collection (stated in Article 3.5), and water sensors confirm the desired water quantity (Fig. 3.9, Chap. 3). Advancement on sensors application and remotely handling is mostly under research to ensure the desired water quality and quantity.

8.7 Economic and Financial Analysis

The economic analysis estimates economic, environmental, and social effects. Here, the criteria of economic feasibility are the ratio of benefits- costs (B/C). If the B/C is greater than 1, the designed harvesting project is economically feasible. If the B/C is less than 1, the project is not economically viable. The detailed consideration should be:

Cost

- The economic cost is the initial investment, including operation and maintenance costs.
- The environmental cost is usually absent for rainwater harvesting. However, consideration should be taken in terms of altering the hydrological cycle.
- Social cost considers the harvested water quality issues, i.e., risk due to (i) rainwater as drinking, (ii) rainwater and stormwater as irrigation, (iii) combined rainwater, stormwater, and greywater as irrigation or MAR.

Benefit

- The economic benefit is the availability of water in the dry period.
- The environmental benefits are the water and energy savings.
- Social benefits are time-saving for securing the required amount of water, reducing urban floods, and improving urban lifestyle.

The financial analysis evaluates financial cost and financial benefits. Here, the criteria of financial feasibility are whether the designed harvesting is financially attractive than the existing practices for surface or groundwater. The NPV for a different mode of water harvesting options is calculated. Both of these analyses could be modeled using the RainCycle.

8.7.1 Reliability Analysis

A reliability model for the rainwater harvesting system depends mainly on rainfall data availability. The reliability of a rainwater harvesting system has been expressed as a time or volumetric basis. *Volumetric reliability* is the total volume of harvested water supplied divided by the total desired water demand, also known as water-saving efficiency, i. e., the fraction of runoff from the contributing catchment utilized. Volumetric water-saving efficiency (E_{ws}) formula:

$$E_{ws} = \frac{\sum_{i=1}^{n} Y_i}{\sum_{i=1}^{n} D_i} \times 100 \tag{8.1}$$

where, n is the total time intervals in simulation, Y is the rainfall volume yielded to meet the water demand, D is the water demand.

Low water-saving efficiency refers to the lesser rainwater yield than the demand. If the water-saving efficiency is high, the sufficient rainfall amount to meet the demand, and the user can consider increasing the water demand for other usages.

Time reliability is the total harvestable water volume divided by the time to meet the desired demand. Thus, the reliability of a storage tank has been described by Imteaz et al. (2021) as:

$$R_e = \frac{N - U}{N} \times 100 \tag{8.2}$$

Here, Re is the tank's reliability to supply intended demand (%), U is the number of days the tank cannot meet the demand, and N is the total number of days serves in a particular year.

Performance evaluation using time reliability is rational for the priority-based systems designed for drinking water in developing and developed countries. However, urban water supply for non-potable works under the top-up function, and volumetric reliability applies to evaluate performances. Therefore, hydrological performance relates to the *volumetric reliability* of the rainwater harvesting system than in the absence of proper standards.

8.7.2 Statistical Analysis

A series of statistical analyses are described in Chap. 2 on rainfall analysis. Statistical analysis tools play a vital role in this regard, i.e., Statistical Package Social Science (SPSS), R, etc. Probability computation plays a crucial role in this regard (are described in Chap. 2). A nonparametric stochastic rainfall generator often requires future prediction to a planned rainwater harvesting.

The water balance model simulates the inflow and outflow of rainwater from the storage within a selected time duration. The daily water balance model considers daily rainfall, catchment, losses, spillage, evaporation, storage (tank) volume, and water uses.

8.8 Assessment of Socio-Economic Impacts

The success of adopting rainwater harvesting largely depends on socio-economical awareness and the performance of the system. As stated in the earlier section on economic analysis, consideration is required for social and environmental. Due to rainwater harvesting, different environmental practices develop more targeted social marketing strategies to facilitate behaviour change to obtain a water smart city. The desirable behavioural changes are users' awareness (education programs), application (prescriptive requirements in planning codes), or acquisition (capacity building) regarding rainwater harvesting. The UK's strategic socio-framework for rainwater harvesting transition comprises actor, action, and aim Ward et al. (2010). Institution/service providers and end-user are the actors, and their activity includes capacity development, support services, and product development. And the aim has an institutional commitment, social receptivity, and technical relevance.

The hybrid application of GIS and Multi-Criteria Decision Analysis (MCDA) has been frequently used for site selection and management of rainwater harvesting systems (Singh et al. 2004; Chowdhury 2014; Jha et al. 2014; Ejegu and Yegizaw 2020). GIS-based MCDA offers an excellent basis for the decision support system to plan, design, and execute a rainwater harvesting system. Thus, a potential site identification for rainwater harvesting incorporates socioeconomic factors; MCDA includes distance from drainage networks, physical properties of the catchment area, and rainfall trends. MCDA for surface runoff reduction and managed aquifer recharge should consider socioeconomic to other factors, i.e., rainfall patterns, LULC, wells' positions, drainage network, and physical properties of the catchments.

8.8.1 *Analytic Hierarchy Process (AHP)*

The Analytic Hierarchy Process (AHP) organizes and analyzes complex decisions by arithmetic and respondent's thinking. Thomas L. Saaty has developed AHP in the 1970s, containing three parts: the ultimate goal, possible alternatives, and the judging criteria to the alternative options. AHP offers a realistic framework for a critical judgment by quantifying measures and different opportunities and linking them to the overall goal. The AHP is an algorithm involving pairwise comparison matrices that can be included in MCDA. Thus, stakeholders can compare desirable two criteria at a time through pair-wise comparisons. AHP converts these evaluations into numbers, and this allows to reach all the feasible measures. This quantifying capability differentiates the AHP from the available decision-making techniques. Finally, arithmetical significances are calculated from the different options. These could signify the preferred way out based on all the respondents' choices.

AHP has been applied to identify rainwater harvesting potential zones to secure alternative water supply in the city, reduce surface runoff, and managed aquifer recharge (Mohanty et al. 2013; Akter and Ahmed 2015; Akter et al. 2020). Additionally, the AHP ensures the effectiveness of geospatial data on the Decision Support Systems.

8.8.2 Agent-Based Model (ABM)

Agent-based modeling (ABM) is a promising approach and reasonably advancing the performances in the last two decades. Social scientists have adopted agent-based models in urban and geospatial studies to frame complex and dynamic processes effectively. Several attempts also define the 'Agent'. The most known definition is: an *agent* is a computer system located in various flexible and autonomous environments to meet its design objectives) (Jennings et al. 1999).

There are two properties of agents, viz. autonomy and social ability. In urban planning, ABM has been applied to simulate: geospatial and social science, economics, ecology, environment, and transportation systems. On the other hand, in architectural design, ABMs are typically used to maintain synchronous assistance, permitting multiple users from multi-disciplinary to operate real-time data, i.e., the *collaborative method*. The adoption of rainwater tanks was decided by selecting the behaviour of two types of agents, i.e., regulatory agent and household agent (Castonguay et al. 2018). The ABM approach facilitates connecting sub-models of households' water demand, rain storage's water balance, and households' decision-making.

8.9 Evaluation of Systems

Evaluation of both existing and proposed rainwater harvesting systems needs to be carried out beforehand. This could be conducted through performance analysis, life cycle assessment, analytical probabilistic modeling, and behavioural modeling.

In *Performance Analysis*, a system's performance examines to support optimal design under various rainfall regimes and other factors. Regressional analysis usually carries for developing climate scenarios to design storage. Performance analysis for identifying the storage capacity has been carried out using various models, including analytical probabilistic and behavioural models.

Life Cycle Assessment (LCA) of a rainwater harvesting system has been widely used in diverse sectors since its inception in the late 1960s to provide a quantitative environmental impacts analysis. The LCA also uses all three categories to establish a water smart city, i.e., resource depletion, ecosystem, and human health. Then, LCA also considers energy assessment and economic feasibility assessment.

The *Probabilistic model* for rainwater harvesting system uses mass balance equations for every single component in the rainwater harvesting system, i.e., catchment, conveyance, rainwater storage, and infiltration facilities. Thus, this model would be helpful to identify water savings as well as the impacts of adopting a rainwater harvesting system.

The Behavioural model determines the savings capability for potable water by ensuring the total volume of rainwater consumed and the required potable water. The Neptune (Ghisi 2010) is a computer simulation program; Neptune estimates the captured volume of rainwater. Then, the program calculates the available rainwater in the storage and the daily consumption.

References

Akter A, Ahmed S (2015) Potentiality of rainwater harvesting for an urban community in Bangladesh. J Hydrol 528. https://doi.org/10.1016/j.jhydrol.2015.06.017

Akter A et al (2020) Predicting groundwater recharge potential zones using geospatial technique. Sustain Water Resour Manag 24(6). https://doi.org/10.1007/s40899-020-00384-w

Alan M, Ali F (2011) Comparison of rainfall interpolation methods in a mountainous region of a Tropical Island. J Hydrol Eng 16(4):371–383, American Society of Civil Engineers. https://doi.org/10.1061/(ASCE)HE.1943-5584.0000330

Berne A et al (2004) Temporal and spatial resolution of rainfall measurements required for urban hydrology. J Hydrol 299(3):166–179. https://doi.org/10.1016/j.jhydrol.2004.08.002

Castonguay AC et al (2018) Modelling urban water management transitions: a case of rainwater harvesting. Environ Model Softw 105:270–285. https://doi.org/10.1016/j.envsoft.2018.05.001

Chen Y et al (2016) Integrated design workflow and a new tool for urban rainwater management. J Environ Manage 180:45–51. https://doi.org/10.1016/j.jenvman.2016.04.059

Cheng WYY et al (2009) Influence of cloud condensation and giant cloud condensation nuclei on the development of precipitating trade wind cumuli in a large eddy simulation. J Geophys Res Atmos 114(D8), Wiley. https://doi.org/10.1029/2008JD011011

Chowdhury PA (2014) Future growth trend & potential residential area identification of a city: a case study of chittagong. Curr Urban Stud 02(03):168–177. https://doi.org/10.4236/cus.2014.23017

Dixon A et al (1999) Water saving potential of domestic water reuse systems using greywater and rainwater in combination. Water Sci Technol 39(5):25–32. https://doi.org/10.2166/wst.1999.0218

Ejegu MA, Yegizaw ES (2020) Potential rainwater harvesting suitable land selection and management by using GIS with MCDA in Ebenat District, Northwestern Ethiopia. J Degrad Min Land Manag 8(1). https://doi.org/10.15243/jdmlm.2020.081.2537. https://jdmlm.ub.ac.id/index.php/jdmlm/article/view/762

Fabry F et al (1994) High resolution rainfall measurements by radar for very small basins: the sampling problem reexamined. J Hydrol 161(1):415–428. https://doi.org/10.1016/0022-1694(94)90138-4

Fewkes A (2000) Modelling the performance of rainwater collection systems: towards a generalised approach. Urban Water 1(4):323–333. https://doi.org/10.1016/S1462-0758(00)00026-1

Friedman I et al (1991) Isotopic changes during snow metamorphism, in stable isotope geochemistry. In: Taylor Jr HP, O'Neil JR, Kaplan IR (eds.) A tribute to Samuel Epstein, Geochem. Soc. Spec. Publ. 3, University Park, Penn, pp 211–221

Ghisi E (2010) Parameters influencing the sizing of rainwater tanks for use in houses. Water Resour Manage 24(10):2381–2403. https://doi.org/10.1007/s11269-009-9557-4

Harris D et al (2001) Multiscale statistical properties of a high-resolution precipitation forecast. J Hydrometeorol 2(4):406–418, American Meteorological Society, Boston MA, USA. https://doi.org/10.1175/1525-7541(2001)002<0406:MSPOAH>2.0.CO;2

Herrmann T, Schmida U (2000) Rainwater utilisation in Germany: efficiency, dimensioning, hydraulic and environmental aspects. Urban Water 1(4):307–316. https://doi.org/10.1016/S1462-0758(00)00024-8

Huang G et al (2010) Strengthening of tropical Indian Ocean teleconnection to the northwest Pacific since the mid-1970s: an atmospheric GCM study. J Climate 23:5294–5304

Hutchinson MF (1988) Interpolation of rainfall data with thin plate smoothing splines. Part II: analysis of topographic dependence, pp 168–185

Imteaz MA et al (2021) Impacts of climate change on weather and spatial variabilities of potential water savings from rainwater tanks. J Clean Prod 311:127491. https://doi.org/10.1016/j.jclepro.2021.127491

Jennings S et al (1999) Structural change in an exploited fish community: a consequence of differential fishing effects on species with contrasting life histories. J Animal Ecol 68(3):617–627, Wiley. https://doi.org/10.1046/j.1365-2656.1999.00312.x

Jha MK et al (2014) Rainwater harvesting planning using geospatial techniques and multicriteria decision analysis. Resour Conserv Recycl 83:96–111. https://doi.org/10.1016/j.resconrec.2013.12.003

Kim Y, Han M (2006) A rainfall-storage-runoff (RSR) model for the design of a rainwater tank effective for flow control in urban drainage pipes. In: Proceedings of the 2nd IWA international rainwater harvesting workshop. Beijing, China, 11 September 2006. IWA

Krishnamurthy A et al (2000) Groundwater resources development in hard rock terrain—an approach using remote sensing and GIS techniques. Int J Appl Earth Obs Geoinf 2(3–4):204–215

Mohanty S et al (2013) 'Comparative evaluation of numerical model and artificial neural network for simulating groundwater flow in Kathajodi-Surua Inter-basin of Odisha, India. J Hydrol 495:38–51. https://doi.org/10.1016/j.jhydrol.2013.04.041

Morin J et al (1995) Radar rain field evaluation and possible use of its high temporal and spatial resolution for hydrological purposes. J Hydrol 172(1):275–292. https://doi.org/10.1016/0022-1694(95)02700-Y

Pokhrel YN et al (2012) Model estimates of sea-level change due to anthropogenic impacts on terrestrial water storage. Nature Geosci 5(6)389–392, Nature Publishing Group

Riad PH et al (2011) Application of the overlay weighted model and Boolean logic to determine the best locations for artificial recharge of groundwater. J Urban Environ Eng 5(2):57–66

Roebuck RM, Ashley RM (2007) Predicting the hydraulic and life-cycle cost performance of rainwater harvesting systems using a computer based modelling tool. Water Pract Technol 2(2). https://doi.org/10.2166/wpt.2007.046

Rossman LA (2006) Storm water management model quality assurance report, dynamic wave flow routing. EPA/600/R-06/097

Scarlett M (2021) Future cities: from Le Corbusier's radiant city to the dutch "breathing city 2050. 07 Feb 2021. ArchDaily. ISSN 0719. https://www.archdaily.com/956458/future-cities-from-le-corbusiers-radiant-city-to-the-dutch-breathing-city-2050. Accessed 22 Jun 2021

Schmugge TJ et al (2002) Remote sensing in hydrology. Adv Water Resour 25(8):1367–1385. https://doi.org/10.1016/S0309-1708(02)00065-9

Sener E et al (2005) An integration of GIS and remote sensing in groundwater investigations: a case study in Burdur, Turkey. Hydrogeol J 13(5):826–834. https://doi.org/10.1007/s10040-004-0378-5

Shaban A et al (2006) Use of remote sensing and GIS to determine recharge potential zones: the case of Occidental Lebanon. Hydrogeol J 14(4):433–443. https://doi.org/10.1007/s10040-005-0437-6

Singh KP et al (2004) Multivariate statistical techniques for the evaluation of spatial and temporal variations in water quality of Gomti River (India)—a case study. Water Res 38(18):3980–3992. https://doi.org/10.1016/j.watres.2004.06.011

Solomon S, Quiel F (2006) Groundwater study using remote sensing and geographic information systems (GIS) in the central highlands of Eritrea. Hydrogeol J 14(6):1029–1041. https://doi.org/10.1007/s10040-006-0096-2

Sun Q et al (2018) A review of global precipitation data sets: data sources, estimation, and intercomparisons. Rev Geophys. 56(1):79–107, Wiley. https://doi.org/10.1002/2017RG000574

Tomczak M (1998) Spatial Interpolation and its uncertainty using automated anisotropic inverse distance weighting (Idw)-cross-validation/Jackknife approach. J Geogr Inf Decis Anal 2(2):18–30

Tweed SO et al (2007) Remote sensing and GIS for mapping groundwater recharge and discharge areas in salinity prone catchments, southeastern Australia. Hydrogeol J 15(1):75–96. https://doi.org/10.1007/s10040-006-0129-x

Vaes G, Berlamont J (2001) The effect of rainwater storage tanks on design storms. Urban Water 3(4):303–307. https://doi.org/10.1016/S1462-0758(01)00044-9

Vieritz AM et al (2015) Chapter 2. Rainwater tank modelling. In: Sharma AK, Begbie D, Gardner T (eds) Rainwater tank systems for urban water supply: design, yield, energy, health risks, economics and social perceptions. IWA Publication, Alliance House, London SW1H oQS

Ward S et al (2010) Harvested rainwater quality: the importance of appropriate design. Water Sci Technol 61(7):1707–1714. https://doi.org/10.2166/wst.2010.102

Wong CL et al (2016) Rainfall characteristics and regionalization in Peninsular Malaysia based on a high resolution gridded data set. Water. https://doi.org/10.3390/w8110500

Appendix A

See Table A.1.

Table A.1 Countrywise annual rainfall classification (based on 1960–2017)

Country name	Low rainfall (0–435 mm) Rainfall in year (mm/year)
Afghanistan	51
Angola	56
Albania	59
United Arab Emirates	74
Argentina	78
Antigua and Barbuda	83
Australia	89
Austria	92
Burundi	111
Belgium	121
Benin	125
Burkina Faso	151
Bangladesh	167
Bulgaria	207
Bahrain	216
Bahamas, The	220
Belize	228
Bolivia	228
Brazil	241
Barbados	252

(continued)

© The Editor(s) (if applicable) and The Author(s), under exclusive license to Springer Nature Switzerland AG 2022

A. Akter, *Rainwater Harvesting—Building a Water Smart City*, Springer Water,
https://doi.org/10.1007/978-3-030-94643-2

Table A.1 (continued)

Country name	Low rainfall (0–435 mm) Rainfall in year (mm/year)
Brunei Darussalam	282
Bhutan	282
Botswana	285
Central African Republic	322
Canada	327
Switzerland	346
Chile	402
China	416
Cote d'Ivoire	416
Cameroon	435
Country name	Medium rainfall (436–934 mm) Rainfall in year (mm/year)
Congo, Dem. Rep	494
Congo, Rep	495
Colombia	498
Comoros	534
Cabo Verde	536
Costa Rica	537
Cuba	560
Cyprus	589
Germany	591
Djibouti	593
Dominica	600
Denmark	608
Dominican Republic	624
Algeria	630
Ecuador	636
Egypt, Arab Rep	637
Spain	645
Finland	652
Fiji	657
France	661
Gabon	686
United Kingdom	700
Ghana	703
Guinea	715

(continued)

Table A.1 (continued)

Country name	Medium rainfall (436–934 mm) Rainfall in year (mm/year)
Gambia, The	748
Guinea-Bissau	758
Equatorial Guinea	778
Greece	788
Grenada	788
Guatemala	832
Guyana	836
Honduras	847
Haiti	854
Hungary	867
Indonesia	900
India	934
Country name	Normal rainfall (935–1543 mm) Rainfall in year (mm/year)
Ireland	1010
Iran, Islamic Rep	1020
Iraq	1030
Iceland	1032
Israel	1039
Italy	1054
Jamaica	1071
Jordan	1083
Japan	1110
Kenya	1118
Cambodia	1130
St. Kitts and Nevis	1146
Korea, Rep	1150
Kuwait	1168
Lao PDR	1180
Lebanon	1181
Liberia	1187
Libya	1212
St. Lucia	1220
Sri Lanka	1274
Lesotho	1274
Luxembourg	1292

(continued)

Table A.1 (continued)

Country name	Normal rainfall (935–1543 mm) Rainfall in year (mm/year)
Morocco	1300
Madagascar	1335
Maldives	1343
Mexico	1348
Mali	1410
Malta	1414
Myanmar	1422
Mongolia	1427
Mozambique	1440
Mauritania	1485
Mauritius	1500
Malawi	1500
Malaysia	1513
Namibia	1522
Niger	1537
Nigeria	1543
Country name	High rainfall (1544–2200 mm) Rainfall in year (mm/year)
Nicaragua	1577
Netherlands	1583
Norway	1604
Nepal	1622
New Zealand	1646
Oman	1651
Pakistan	1668
Panama	1705
Peru	1712
Philippines	1732
Papua New Guinea	1738
Poland	1761
Puerto Rico	1784
Korea, Dem. People's Rep	1821
Portugal	1831
Paraguay	1834
West Bank and Gaza	1904
Qatar	1940

(continued)

Table A.1 (continued)

Country name	High rainfall (1544–2200 mm) Rainfall in year (mm/year)
Romania	1972
Rwanda	1976
Saudi Arabia	1996
Sudan	2041
Senegal	2044
Singapore	2051
Solomon Islands	2054
Sierra Leone	2083
El Salvador	2091
Somalia	2156
Sao Tome and Principe	2200
Country name	Very high rainfall (2201–3240) Rainfall in year (mm/year)
Suriname	2200
Sweden	2274
Swaziland	2280
Seychelles	2301
Syrian Arab Republic	2330
Chad	2331
Togo	2348
Thailand	2350
Timor-Leste	2387
Trinidad and Tobago	2391
Tunisia	2497
Turkey	2526
Tanzania	2592
Uganda	2666
Uruguay	2702
United States	2722
St. Vincent and the Grenadines	2875
Venezuela, RB	2926
Vietnam	2928
Yemen, Rep	3028
South Africa	3142
Zambia	3200
Zimbabwe	3240

Appendix B

See Tables B.1 and B.2.

Table B.1 Hydraulic conductivities

Soil type	Hydraulic conductivity, K (m/d)	
	Min	Max
Fine sand	0.02	17.28
Medium sand	0.08	43.20
Corse sand	0.08	518.40
Sand; clean; good aquifer	0.86	864
Sand/gravelly sand; poorly graded; little to no fines	2.20	46.22
Sand/gravelly sand; well graded; little to no fines	8.64×10^{-04}	0.09
Inorganic silty fine sand/clayey fine sand; slight plasticity	4.32×10^{-04}	0.09
Silty sand	8.64×10^{-04}	0.43
Clayey sand	4.75×10^{-04}	0.48
Alluvial gravel/sand	34.56	345.60
Sand/gravel; uniform	345.60	34,560
Sand/gravel; well graded; no fines	3.46	345.60
Gravel	25.92	2592
Gravel/sandy gravel; well graded; little to no fines	43.20	4320
Gravel/sandy gravel; poorly graded; little to no fines	43.20	4320
Silty gravel/silty sandy gravel	4.32×10^{-3}	0.43
Clayey gravel/clayey sandy gravel	4.32×10^{-4}	0.43
Inorganic silt; high plasticity	8.64×10^{-6}	4.32×10^{-3}
Silt; compacted	6.05×10^{-5}	6.05×10^{-3}
Inorganic clay/silty clay/sandy clay; low plasticity	4.32×10^{-5}	4.32×10^{-3}
Organic clay/silty clay; low plasticity	4.32×10^{-4}	8.64×10^{-3}

(continued)

© The Editor(s) (if applicable) and The Author(s), under exclusive license
to Springer Nature Switzerland AG 2022

A. Akter, *Rainwater Harvesting—Building a Water Smart City*, Springer Water,
https://doi.org/10.1007/978-3-030-94643-2

Table B.1 (continued)

Soil type	Hydraulic conductivity, K (m/d)	
	Min	Max
Marine clay; unweathered	6.91×10^{-8}	1.73×10^{-4}
Organic clay; high plasticity	4.32×10^{-5}	8.64×10^{-3}
Inorganic clay; high plasticity	8.64×10^{-6}	8.64×10^{-3}
Clay	8.64×10^{-7}	4.06×10^{-4}
Clay; compacted	8.64×10^{-6}	8.64×10^{-5}
Limestone/dolomite	8.64×10^{-5}	0.52
Sandstone	2.59×10^{-5}	0.52
Siltstone	8.64×10^{-7}	1.21E-3
Anhydrite	3.46×10^{-8}	1.73E-3
Shale	8.64×10^{-9}	1.73E-4
Permeable basalt	0.03	1728.00
Igneous/metamorphic rock; fractured	6.91×10^{-4}	25.92
Granite; weathered	0.29	4.49
Gabbro; weathered	0.05	0.33
Basalt	1.73×10^{-6}	0.04
Igneous/metamorphic rock; unfractured	2.59×10^{-9}	1.73×10^{-5}

Table B.2 Manning's roughness coefficients, n (Chow 2009)

	Manning's n
I. Closed conduits	
A. concrete pipe	0.011–0.013
B. corrugated-metal pipe or pipe-arch	
1. Riveted pipe:	
a. Plain or fully coated	0.024
b. Paved invert (range values are for 25 and 50% of circumference paved)	
(1) Flow full depth	0.021–0.018
(2) Flow 0.8 depth	0.021–0.016
(3) Flow 0.6 depth	0.019–0.013
2. Field bolted:	0.03
C. Vitrified clay pipe	0.012–0.014
D. Cast-iron pipe, uncoated	0.013
E. Steel pipe	0.009–0.011
F. Brick	0.014–0.017
G. Monolithic concrete:	

(continued)

Table B.2 (continued)

	Manning's n
1. Wood forms, rough	0.015–0.017
2. Wood forms, smooth	0.012–0.014
3. Steel forms	0.012–0.013
H. Cemented rubble masonry walls:	
1. Concrete floor and top	0.017–0.022
2. Natural floor	0.019–0.025
I. Laminated treated wood	0.015–0.017
J. Vitrified clay liner plates	0.015
II. Open channels, lined (straight alightment):	
A. Concrete with surfaces as indicated:	
1. Formed, no finish	0.013–0.017
2. Trowel finish	0.012–0.014
3. Float finish	0.013–0.015
4. Float finish, some gravel on bottom	0.015–0.017
5. Gunite, good section	0.016–0.019
6. Gunite, wavy section	0.018–0.022
B. Concrete, bottom float finish, sides are as indicated:	
1. Dressed stone in mortar	0.015–0.017
2. Random stone in mortar	0.017–0.020
3. Cement rubble masonry	0.020–0.025
4. Cement rubble masonry, plastered	0.016–0.020
5. Dry rubbel (riprap)	0.020–0.030
C. Gravel bottom, sides as indicated	
1. Formed concrete	0.017–0.020
2. Random stone in mortar	0.020–0.023
3. Dry rubble (riprap)	0.023–0.033
D. Brick	0.014–0.017
E. Asphalt:	
1. Smooth	0.013
2. Rough	0.016
F. Wood, planed, clean	0.011–0.013
G. Concrete-lined excavated rock:	
1. Good section	0.017–0.020
2. Irregular section	0.022–0.027
III. Pen channels, excavated (straight alignment, natural lining):	
A. Earth, uniform section:	
1. Clean, recently completed	0.016–0.018

(continued)

Table B.2 (continued)

	Manning's n
2. Clean, after weathering	0.018–0.020
3. With short grass, few weeds	0.022–0.027
4. In gravely soil, uniform section, clean	0.022–0.025
B. Earth, fairly uniform section:	
1. No vegetation	0.022–0.025
2. Grass, some weeds	0.025–0.030
3. Dense weeds or aquatic plants in deep channels	0.030–0.035
4. Sides clean, gravel bottom	0.025–0.030
5. Sides clean, cobble bottom	0.030–0.040
C. Dragline excavated or dredged:	
1. No vegetation	0.028–0.033
2. Light brush on banks	0.035–0.050
D. Rock:	
1. Based on design section	0.035
2. Based on actual mean section:	
a. smooth and uniform	0.035–0.040
b. Jagged and irregular	0.040–0.045
E. Channels not maintained, weeds and brush uncut:	
1. Dense weeds, high as flow depth	0.08–0.12
2. Clean bottom, brush on sides	0.05–0.08
3. Clean bottom, brush on sides, highest stage of flow	0.07–0.11
4. Dense brush, high stage	0.10–0.14
IV. Highway channels and swales with maintained vegetation (for velocities of 0.61–1.83 m/s)	
A. Depth of flow upto 213 mm	
1. Bermudagrass, Kentucky bluegrass, Buffalograss:	
a. Mowed to 51 mm	0.07–0.045
b. Length 101–152	0.09–0.05
2. Good stand, any grass:	
a. Length about 305 mm	0.18–0.09
b. Length about 610 mm	0.30–0.15
3. Fair stand, any grass:	
a. Length about 305 mm	0.14–0.08
b. Length about 610 mm	0.25–0.13
B. Depth of flow 213–457 mm	
1. Bermudagrass, Kentucky bluegrass, Buffalograss:	
a. Mowed to 51 mm	0.05–0.035

(continued)

Table B.2 (continued)

	Manning's n
b. Length 101–152	0.06–0.04
2. Good stand, any grass:	
a. Length about 305 mm	0.12–0.07
b. Length about 610 mm	0.20–0.10
3. Fair stand, any grass:	
a. Length about 305 mm	0.10–0.06
b. Length about 610 mm	0.17–0.09
V. Street and expressway gutters:	
A. Concrete gutter, troweled finish	0.012
B. Asphalt pavement:	
1. Smooth texture	0.013
2. Rough texture	0.016
C. Concrete gutter with asphalt pavement:	
1. Smooth	0.013
2. Rough	0.015
D. Concrete gutter with asphalt pavement:	
1. Smooth	0.014
2. Rough	0.016
E. For gutters with small slope, where sediment may accumulate, increase above value of n by	0.002
VI. Natural stream channels:	
A. Minor streams (surface width at flood stage less than 305 m):	
1. Fairly regular section:	
a. Some grass and weeds, little or no brush	0.030–0.035
b. Dense growth of weeds, depth of flow materially greater than weed height	0.035–0.05
c. Some weeds, light brush on banks	0.035–0.06
d. Some weeds, heavy brush on banks	0.05–0.07
e. Some weeds, dense willows on banks	0.06–0.08
f. For trees within the channel, with branches submerged at a high stage, increase all above values by	0.01–0.02
2. Irregular sections, with pools, slights channel meander, increase values given in 1a to e about	0.01–0.02
3. Mountain streams, no vegetation in the channel, banks usually steep, trees and brush along banks submearged at high stage:	
a. Bottom of gravel, cobblers, and few boulders	0.04–0.05
b. Bottom of cobblers with large boulders	0.05–0.07
B. Floodplains (adjacent to natural streams):	
1. Pasture, no brush:	
a. Shortgrass	0.030–0.035

(continued)

Table B.2 (continued)

	Manning's n
b. High grass	0.035–0.05
2. Cultivated areas:	
a. No crop	0.03–0.04
b. Mature row crops	0.035–0.045
c. Mature field crops	0.04–0.05
3. Heavy weeds, scattered brush	0.05–0.07
4. Light brush and trees:	
a. Winter	0.05–0.06
b. Summer	0.06–0.08
5. Medium to dense brush:	
a. Winter	0.07–0.11
b. Summer	0.10–0.16
6. Dense willows, summer, not bent over by current	0.15–0.2
7. Cleared land with tree stumps	
a. No sprouts	0.04–0.05
b. With heavy growth of sprouts	0.06–0.08
8. Heavy stand of timber, a few down trees, little undergrowth	
a. Flood depth below branches	0.10–0.12
b. Flood depth reaches branches	0.12–0.16
C. Large stream of the most regular section, with no boulders or brush	0.028–0.033

Appendix C

See Table C.1.

© The Editor(s) (if applicable) and The Author(s), under exclusive license
to Springer Nature Switzerland AG 2022

A. Akter, *Rainwater Harvesting—Building a Water Smart City*, Springer Water,
https://doi.org/10.1007/978-3-030-94643-2

Table C.1 Rain garden plants

Common name		Scientific name	Habit	Sunlight and aspect	Origin
Guelder rose		Viburnum opulus	Perennial shurb	Any	Native to Europe, northern Africa and central Asia
Dogwood		Cornus sanguinea	Perennial shurb	Any	Native to most of Europe and western Asia, from England and central Scotland east to the Caspian Sea
Culver's root		Veronicastrum virginicum,	Herbaceous perennial	Full sun or partial shade	Native to the eastern United States and south-eastern Canada
Aster		Aster spp.	Herbaceous perennial	Full sun or partial shade	United Kingdom, Eurasia and North America
Black eye susan		Rudbeckia hirta	Herbaceous annual or biennial	Full sun or partial shade	Native to Eastern and Central North America and naturalized in the Western part of the continent as well as in China

(continued)

Table C.1 (continued)

Common name		Scientific name	Habit	Sunlight and aspect	Origin
Stinking hellebore		Helleborus foetidus	Evergreen perennial	Full sun or part shade	England, mountainous regions of Central and Southern Europe, and Asia Minor
Montbretia		Crocosmia	Evergreen or deciduous perennials	Part shade	Southern and eastern Africa, Sudan
Carpet Bugle		Ajuga reptans	Perennials	Part shade	Europe, North America, UK

(continued)

Table C.1 (continued)

Common name		Scientific name	Habit	Sunlight and aspect	Origin
Wild Columbine		Aquilegia canadensis	Herbaceous perennials	Full sun or part shade	Eastern North America
Inula		Inula hookeri	Herbaceous perennial	Partial sun	Europe, Asia, and Africa
Hemp-agrimonyor Holy rope		Eupatorium cannabinum	Herbaceous perennial	Full or partial sun	China, US and Canada

(continued)

Table C.1 (continued)

Common name		Scientific name	Habit	Sunlight and aspect	Origin
Clustered bellflower or Dane's blood		Campanula glomerata	Herbaceous perennial	Full or partial sun	Eurasia, UK, Japan, Europe North America
Sneezeweed		Helenium autumnale	Herbaceous perennial	Full sun	North America
Periwinkle		Vinca minor	Perennial sub-shrub	Everywhere	Europe

(continued)

Table C.1 (continued)

Common name		Scientific name	Habit	Sunlight and aspect	Origin
Elephant's ears		Bergenia sp.	Rhizomatous perennial	Full or partial sun	Tropical & Subtropical Asia entral Asia, from Afghanistan to China
Plantain lilies		Hosta spp.	Herbaceous perennial	Full or partial shade	Northeast Asia (China, Japan, Korea, and the Russian Far East)
Yellow flag		Iris pseudacorus	Rhizomatous perennial	Full or partial sun	Europe, western Asia and northwest Africa
Siberian flag		Iris sibirica	Rhizomatous perennial	Full or partial sun	Europe and Central Asia (including Armenia, Azerbaijan and Siberia)

(continued)

Table C.1 (continued)

Common name		Scientific name	Habit	Sunlight and aspect	Origin
Garlic and onion		Allium spp.	Bulbous perennial	Full sun	Temperate climates of the Northern Hemisphere
Soft rush		Juncus effusus	Evergreen perennial	Full or partial sun	Eurasia, North America, Australia, New Zealand
Pesndulous sedge		Carex pendula	Rhizomatous perennial	Full or partial sun	Europe, north-west Africa, the Azores, Madeira and parts of the Middle East
Zebra grass		Miscanthus sinensis	Perennial deciduous grass	Full sun	Estern Asia

Reference

Chow VT (2009) Open-channel hydraulics. The Blackburn Press, Caldwell, NJ

Index

© The Editor(s) (if applicable) and The Author(s), under exclusive license to Springer Nature Switzerland AG 2022
A. Akter, *Rainwater Harvesting—Building a Water Smart City*, Springer Water,
https://doi.org/10.1007/978-3-030-94643-2

R

S

T

U

V